Nuclear Power, Pollution and Politics

In the years immediately after the Second World War, the answer to the energy problems of developed western countries appeared to be provided by nuclear power. Since then, however, there have been setbacks, delays and difficulties, with considerable controversy about the manner and need for its application. This book charts the technical development of nuclear power and the growing politicisation of debate about its efficacy. It stresses the need to compare all aspects of power systems in order to derive the best strategies to adopt *now* in order to provide satisfactory power supplies in the future.

The early chapters sketch out the background of possible power systems, concentrating on three groups – nuclear, coal-fired, and the 'developing' systems of nuclear fusion and 'renewables'. Systems such as solar heating and energy conservation, which have only minor effects on the total requirement, are also briefly covered. Special attention is paid to the flows of key isotopes in the various possible nuclear cycles, and to the disposal of nuclear waste. Dr Burton outlines the health hazards which may result from power systems, and compares the costs of various systems, including those to human health, flora, fauna and artefacts. He describes the various organisations engaged in the development of power systems, and those opposed to them, and analyses their relationship with the media. In conclusion he summarises the technical and political factors, and proposes an overall power arrangement which offers a balance between these elements.

Essential reading for those working in the nuclear power business, *Nuclear Power, Pollution and Politics* will also be of great interest to planners at national and local level, and to students of economics, politics, geography and environmental studies.

Bob Burton is Managing Director of Nuclear Technology (Consultants). From 1956 to 1978 he worked for UKAEA, latterly as Fuel Cycle and Wastes Manager, and from 1978 to 1980 he was Head of an Environmental Protection Group at BNFL.

NUCLEAR POWER, POLLUTION AND POLITICS

Bob Burton

Routledge

First published 1990 by Routledge
11 New Fetter Lane, London EC4P 4EE

Printed and bound in Great Britain by
Mackays of Chatham PLC, Chatham, Kent

British Library Cataloguing in Publication Data

Burton, Bob, *1927–*
Nuclear power, pollution and politics.
1. Great Britain. Nuclear power
I. Title
621.48'0941

ISBN 0-415-03065-X

CONTENTS

FIGURES

TABLES

PREFACE

A few decades ago, Professor Northcote Parkinson published a series of essays on the science of public and business administration in 'Parkinson's Law or The Pursuit Of Progress' (Parkinson, 1958). The basic form of Parkinson's Law is that 'work expands so as to fill the time available for its completion'. A sequel to this is his Law of Triviality (related to finance committees) - 'the time spent on any item of the agenda will be inversely proportional to the sum (of money) involved'. An example of the latter law in the above book describes the perfunctory discussion on an item concerning an expensive and technically complex nuclear reactor and compares this with the fierce debate over the proposed new bicycle shed for the clerical staff. Though well out of depth on a matter of high technology, no self-respecting committee member would confess to not understanding the construction and economics of a bicycle shed!

Few better illustrations of the workings of the various laws enunciated by Parkinson could be provided than the development of nuclear power and its replacement of fossil-fuelled power. In the decade or so after the Second World War, the future of nuclear power looked distinctly rosy. The basic features of operation had been demonstrated on reactors built for military purposes and, in principle, it was only necessary to abstract heat from a fluid that cooled the reactor, in order to raise steam and operate turbines for the production of electricity. Prospects looked good in the 1950s, as increasingly the debate over the environmental consequences of emissions from coal-fired power and the rise in oil prices lent support to a clean nuclear system of low cost. Politically, too, nuclear power was an attractive prospect, since it promised independence from oil and also a breaking up of the monopoly position of coal and the union dominance in that industry. In the UK, with the prospect of a cheap source of power, forecasts of the level of power demand by the end of the century soared to heights several times that then being supplied. Though there had been an accident in a nuclear reactor at Windscale in Cumbria in 1957, this, like the London 'smog' of 1952, which was partly caused by stack discharges from coal-fired power and killed thousands in a few days, was regarded by the public as a one-off event which would not happen again.

By the early 1970s, however, several issues had emerged which became a focus for pressure groups, who, for various reasons, were opposed to nuclear power. Spent fuels from reactors were being reprocessed to recover plutonium and uranium. The most radioactive part of the wastes from this reprocessing, High Level Wastes (to be described later) were being stored at Windscale, apparently without any established route for disposal. Plutonium was being accumulated in tonnage quantities, equivalent to thousands of atomic bombs; possibilities of theft by terrorists or misuse by foreign countries with their own reactors became a matter of international concern. The media in the UK seized on these issues and vied with each other to maintain an atmosphere of sensation, with little thought as to what was best in the national interest. In line with the above 'Law of Triviality', the pressure groups and media paid special attention to the topic they felt they understood - the disposal of nuclear waste. As research to find satisfactory routes for the treatment and disposal of radioactive wastes progressed (expanding to fill the apparently endless time available, in accordance with Parkinson's Law), the pressure groups continually 'moved the goalposts' to prove that any proposal was inadequate and that, as a consequence, nuclear power should be abandoned. On the other hand, successive UK governments of various political shades supported a steady increase in nuclear power.

Since the accident at Three Mile Island in 1979 and the more serious one at Chernobyl in 1986, the call for a phasing-out of nuclear power has become more insistent. However, if a run-down of nuclear power is contemplated, then the problem of how to replace it and what to do with residual radioactive materials and wastes is exacerbated. Stored plutonium and spent fuel, which contains both the hazardous components of High Level Wastes and plutonium, will present severe radiological problems on disposal, since the toxic potential of plutonium therein is much greater than that of the long-lived radioactivity in High Level Waste. If the plutonium cannot be disposed of conventionally, e.g. by burial, then it can only be destroyed in nuclear reactors - somewhat difficult if these have been closed down! Moreover, alternative power systems, especially that based on coal, also have shortcomings, a factor often carefully ignored by 'Greens' and the media. Inevitably, electricity cannot be produced without some corresponding detriment or social impact, i.e. we cannot have 'something for nothing'! Pollution results from the discharges of the wastes from power production and changes of technology can have considerable effects on the associated workforce. Markedly different stances are then taken up by political parties - hence our title covering all three aspects stemming from the first - Nuclear Power, Pollution and Politics!

It is not the intention of this book to present yet again the familiar postures of the various protagonists in debates on sources of electricity, but to pose the correct questions carefully avoided in many discussions to date, i.e. 'What are the best strategies to adopt and the action to take <u>now</u> to provide satisfactory power supplies in the future?'. A sufficient description of possible UK power processes is therefore provided to allow readers to form their own judgements and perhaps to insist, as consumers and 'paymasters' of electricity, on applying The Golden Rule - 'He Who Holds The Gold Makes The Rules'. To this end, the main text has been set out simply, with technical detail assembled in appendices at the end of each chapter. Special emphasis has been placed on the disposal of nuclear wastes, since these have been the subject of many misleading articles in the media to date. Less emphasis is placed on the detailed 'internal workings' of nuclear reactors and fuel cycle plants. These are 'black boxes': as with motor car engines, most people are concerned about the chances of breakdown and its consequences, not the causes of failure.

Though the circumstances of power production are discussed in terms of the particular conditions prevailing in the UK, similar analyses with different emphases can be deduced for other countries. An important factor in all cases, discussed in the latter part of this book, is the increasing influence of the media, as television has become more pervasive and influential on public opinion. This factor then depends markedly on the degree of control exercised over the media by the State.

Our presentation covers a wide field and is therefore only a 'broad brush' approach: nevertheless, experts in various subjects have been consulted and it is appropriate to express appreciation to them:

Allott and Lomax, Consulting Engineers, Sale, Manchester.

Professor J.H. Fremlin, formerly Professor of Applied Radioactivity, University of Birmingham.

Dr. C.J. Haslam, Senior Lecturer in Economics, Polytechnic of East London.

Dr. G.D. Nicholls, formerly Reader in Geochemistry, University of Manchester, and

Professor K.R. Rushton, Civil Engineering Department, University of Birmingham.

Thanks are also due to numerous colleagues and friends who kindly consented to review the text to permit ready comprehension of technical aspects by the general reader.

INTRODUCTION

In the discussions to follow, possible future power systems have been listed in three groups. Firstly, considerable attention is paid to power derived from nuclear fission (for convenience referred to simply as nuclear power). Secondly, power systems based on coal are discussed, since they are well established and are the logical 'conventional' alternative to nuclear power. Power from oil and gas is not considered; there is a general consensus that supplies of both these energy sources will be scarce in the long term and should be reserved for more sophisticated uses than simply being burnt to produce power. It is widely agreed, too, that further expansion of hydroelectricity is difficult in the UK, because of geographical limitations. Thirdly, systems providing power from nuclear fusion or 'renewable' natural sources are described. Fusion can offer little possibility currently of early large-scale commercial application, though the renewables have the potential to provide a significant component of the UK power mix at the turn of the century. It is therefore important to know whether it is worth waiting for their development and so include them in long-term planning. Processes which do not produce power as such, but could reduce its requirement, e.g. solar heating, conservation of energy and the use of waste heat from power stations, are introduced briefly in early chapters, with their overall effect on future power demand assessed towards the end of the book.

Chapter 1 sets out briefly the technical background of possible power systems in sufficient detail for the reader to follow later comparisons.

Chapter 2 presents the flows of key nuclides in nuclear fuel cycles.

Chapter 3 discusses the wastes that arise from each type of power system.

Chapter 4 describes the disposal of liquid and solid forms of such wastes by methods used at present and probable ones in the future.

Chapter 5 introduces the types of health hazards which ensue from operating power systems.

Chapter 6 sets out broad estimates of the casualties associated with routine discharges from the various power systems.

Chapter 7 reviews large-scale accidents which have occurred in the operation of the various systems since the Second World War.

Chapter 8 explores the costs of power systems and their interaction with competing financial requirements, e.g. the National Health Service.

Chapter 9 looks at the views of various technical organisations, political parties and pressure groups involved in UK power.

Chapter 10 summarises the technical and political factors of preceding chapters and puts forward an overall power arrangement which offers a balance between these factors. Supporting technical demonstrations and regulatory changes are also suggested.

For the reader's convenience, a glossary has been appended of technical and organisational terms arising throughout the book.

Chapter One

THE BASICS OF POWER SYSTEMS

1.1 GENERAL

In the UK and most developed countries, a large proportion of power supplies is derived through the creation of electricity which is then distributed to consumers via a nationwide transmission grid. The electricity is produced in large-scale units by rotating a generator shaft (the rotor) within a stationary unit (the stator), so that the relative motion of the magnetic field and electric windings on the rotor and stator causes currents to flow in the windings. The mechanical energy of the rotor is provided by a flow of fluid against the blades of the turbine. This flow may be available naturally, as in wind-driven propellors, or as pressurised water in hydroelectric schemes; on the other hand, the energy form available may be unsuitable for such a direct conversion to electricity. In the case of coal-fired power, for example, the chemical energy from the reaction of coal with oxygen is converted to heat by burning in air, the heat then being used to boil water. The resulting steam flows against turbine blades to create power as described above. After passage through the turbine, the steam is condensed by flowing through metal tubes cooled by water. The latter can then be discharged to waste if there is an ample water supply, as in the sea, or cooled by air in large cooling towers and reused in the condenser.

Of the power systems to be described, coal-fired power and nuclear power require conversion of their basic energy to heat so as to drive a turbine by steam. Other systems, such as wind or wave machines or tidal energy schemes, can convert their energy to power without a heat production stage.

In this chapter, an outline of the technology of possible power systems for the UK is provided sufficient for the reader to follow the comparisons developed in later chapters. Subsidiary aspects, such as the arisings of wastes and their disposal, are described in Chapters Three and Four.

1.2 NUCLEAR POWER

1.2.1 Energy from Nuclear Reactions

In simple terms, each of the atoms of which matter is composed has the preponderance of its mass concentrated in a nucleus occupying only a tiny fraction of the atomic volume. This nucleus has a positive electric charge balanced by a number of electrons round it, each bearing unit negative charge. Interactions of these electrons with those from other atoms determine everyday chemical reactions. The nucleus itself can be considered to contain units of effectively equal mass - protons and neutrons. The proton has a positive electrical charge equal and opposite to that on an electron, whereas the neutron has no charge. Consequently, an atomic number may be defined, which is the number of positive charges or protons in the nucleus or the number of electrons in the atom. Each chemical element, which is composed of atoms with the same number of electrons, is then defined by its atomic number, e.g. helium has an atomic number of 2.

The mass of an atom is determined approximately by the sum of the numbers of protons and neutrons in the nucleus, called the mass number. This mass is less than that corresponding to the protons and neutrons as separate units, due to the binding between protons and neutrons. There is a stronger binding per unit mass within nuclei of medium mass number (say 50 to 150) than for lighter or heavier nuclei. This means that, if (say) two light nuclei react to form a nucleus of medium mass, or a heavy nucleus fissions, i.e. splits, to form two nuclei of medium mass, there is some loss of mass. By Einstein's law of equivalence of mass and energy, this loss appears as a considerable amount of energy. These energies are respectively those of nuclear fusion and fission. Examples are the fusion of two deuterium nuclei of mass number 2 to form a helium nucleus of mass number 4, and the fission of uranium of mass number 235 into two parts of medium mass number. The energy of such reactions is very large: the fission of one gram of uranium or fusion of a few milligrams of deuterium yields energy equivalent to burning a tonne (1,000 kilograms) of coal.

Naturally occurring elements have atomic numbers from 1 (hydrogen) to 92 (uranium). However, for some of these elements, more than one number of neutrons can occur in a nucleus of the same charge, i.e. the nuclei have the same atomic number but different mass numbers: such nuclei are known as isotopes of the element. Each 'nuclide' with atoms of a particular nucleus can then be defined by its element and mass number, e.g. uranium-235 or in abbreviated form U235. The manner in which the elements with their various isotopes have been formed and eventually distributed on Earth is outlined in Appendix 1.1.

The numbers of protons and neutrons in nuclei of naturally occurring elements are roughly equal, but, in general, there is an increasing slight excess of neutrons over protons as the mass number increases. This is a very important feature for the operation of a nuclear fission reactor, where heavy nuclei are caused to fission by collision with neutrons. The so-called 'fission products' from a given heavy nucleus have atomic numbers of which one is in the range 30 to 50 and the other in the range 70 to 50. There is no significant loss of protons in the split, but overall there is a lower requirement for neutrons in the product nuclei. Some of the excess of these are emitted as individual neutrons capable of causing further fissions, so providing the condition for a continuous output of fission energy.

1.2.2 **Radioactivity**

If an atomic nucleus is unstable, it can change to another type of nucleus, its daughter, simultaneously emitting radiation, i.e. radioactivity. The unit of radioactivity is the becquerel (Bq) denoting one radioactive decay per second. A multiple of this, often used for convenience in avoiding very large numbers, is the Terabecquerel (TBq), which is one million million Bq. The historical unit, the curie (Ci), is equivalent to 0.037 TBqs. If the daughter nuclide is itself unstable, a sequence known as a decay chain result, as exemplified by the uranium-238 chain described in Appendix 1.2 and listed in Table 1.1 with an illustration in Fig. 1.1 at the end of the chapter. The main types of emissions occurring in nuclear power are:

(a) neutrons, leaving a nucleus of the same charge but one unit less in mass,

(b) alpha particles, which have a helium nucleus, i.e. 2 protons and 2 neutrons, thus leaving a decay product nucleus with lower atomic and mass numbers by 2 and 4 respectively,

(c) beta particles, which are electrons, so that each nucleus of decay product has an increase of unity in atomic number but no significant change in mass, and

(d) gamma radiation. This is electromagnetic radiation (and therefore with no charge or mass) which can be emitted with each of the above types of radiation or separately by a change of energy level in a nucleus, in which case the isotope is unchanged in charge or mass.

Alpha radiation mostly arises from 'actinides', i.e. those elements with atomic number equal to or greater than 85, the atomic number of actinium. This type of radiation can be stopped easily, e.g. by a sheet of paper. Beta radiation occurs with some actinides and with many of the fission products. It is usually more penetrating than alpha radiation, requiring, for example, a thin sheet of metal to stop it. Gamma and neutron radiation, arising during fission, fusion and radioactive decay, can be highly penetrating, often needing about one metre or more of concrete to protect the human body.

Apart from radioactivity which arises naturally from some elements in the Earth's surface, the Earth is being constantly bombarded with cosmic rays - high-energy electromagnetic radiation from space. Human beings are therefore irradiated by both these sources and from the secondary radioactivity the cosmic rays cause on impact with the atmosphere and ground.

1.2.3 <u>Nuclear Reactions in Power Production</u>

In the last section, the concept of fission of heavy nuclides by neutrons was introduced; another common reaction in nuclear power is neutron capture. These can both be illustrated with reference to the most important reactions in nuclear power today (see Fig. 1.2).

Uranium-235 + Neutron ---------> Fission Products + Neutrons (A)

Uranium-238 + Neutron ---------> Plutonium-239 (B)

Plutonium-239 + Neutron -------> Fission products + Neutrons (C)

Plutonium-239 + Neutron -------> Plutonium-240 (D).

Reactions A and C yield several new neutrons; provided the concentration of fissile isotopes in nuclear 'fuel' is high enough, such that one of these neutrons collides to cause another fission, 'criticality' is achieved, i.e. the reactions become self-sustaining and continuous energy production is possible. Reaction B (see Appendix 1.3 and Fig. 1.2 for more details of intermediate nuclides) is known as breeding, in that plutonium-239 is produced, which can in turn be easily fissioned in a later type C reaction. Reaction D, or a sequence of similar reactions, uses up neutrons in forming heavy nuclides such as plutonium-240, 241 and 242. Reaction B can be followed successively by C or D in the same reactor, e.g. uranium-fuelled reactors can create plutonium then fission part of it before the fuel is discharged. After such a discharge from a reactor, this 'spent fuel' can be reprocessed chemically to discard fission products and elements such as americium. The uranium and plutonium, which are by contrast useful in a nuclear reactor, can be incorporated in new fuel to provide further fissions in a later reactor cycle.

Two general types of reactor exist, Thermal and Fast Reactors, according to whether the neutrons are deliberately moderated, i.e. slowed down by interspersing light atoms such as hydrogen near the fuel to reduce neutron speeds by elastic collisions, or whether there is no 'moderator'. In this second case, the neutron speeds are fast and so the fissile isotopes find it more difficult to react with them; the fissile isotope concentration must therefore be higher in a Fast Reactor. On the other hand, because of the need to intersperse the fuel with a neutron moderator, the energy-producing core of a Thermal Reactor has a much bigger volume than that of a Fast Reactor. A further important difference between the reactors is that the chance of fission of a heavy isotope (as in reaction C) relative to neutron capture (as in reaction D) is greater in a Fast Reactor. A Fast Reactor thus avoids a problem associated with Thermal Reactors, where successive recycling of plutonium through reactors and reprocessing plants leads to a lower 'quality' of plutonium, i.e. an increase in its heavier isotopes by reactions of type D, particularly plutonium-240 and 242, which fission with difficulty in Thermal Reactors. The reactivity of fuel and its control in reactors is discussed in Appendix 1.4.

Finally, it is important to mention here one more type of radioactivity: this is induced by capture of neutrons in structural parts of a nuclear reactor, as in the reaction

Cobalt-59 + Neutron ------> Cobalt-60 (E).

Here, the stable isotope of cobalt, an essential component of stainless steel, is converted to the isotope cobalt-60; this decays with a half-life of 5.3 years, emitting high-energy gammas in conjunction with beta particles. Induced cobalt-60 activity is an important feature in nuclear waste, since much nuclear plant equipment contains stainless steel which in the course of time is scrapped.

1.2.4 <u>Reactor Types for UK Nuclear Power</u>

1.2.4.1 <u>General</u>. In this section, the basic nuclear principles of the reactor types most important for UK power are introduced. The physical layout of the reactors is not given here; excellent descriptions are published in Cottrell ([illegible] and Patterson (1983). The basic raw fuel [illegible] ear reactors is uranium, which is present [illegible] to a negligible extent in the UK, most of [illegible] imported as oxide in ore from Canada.

1.2.4.2 Thermal Reactors. Although the readily fissile nuclide uranium-235 is present at a concentration of only 0.72%, (the rest is effectively all uranium-238), Thermal Reactors can be designed to operate with natural uranium as fuel. The resulting reactions of uranium-235 and uranium-238 are described in Appendix 1.3 and shown in Fig. 1.2. In the earliest type of power reactor in the UK, metallic natural uranium is enclosed in a 'can', fabricated from an alloy of magnesium and aluminium known as Magnox. These canned fuel 'pins', interspersed in a neutron moderator of graphite, a natural form of carbon, form key features of Magnox reactors, named after the fuel can. The earliest Magnox stations were built at Calder Hall in Cumbria and at Chapel Cross, just across the Scottish border. Though they have supplied power to the National Grid for about 30 years, they were built primarily for the production of plutonium for military purposes, augmenting the supply from the earlier Windscale 'Piles'. A vital difference is that cooling was done by air in the Piles, whereas the Magnox reactors use recirculated carbon dioxide gas which passes on its heat through the walls of metal tubes in 'heat exchangers' to a conventional steam-raising circuit.

More advanced designs of Thermal Reactors operating and proposed for the UK are the Advanced Gas-Cooled Reactor (AGR) and the Pressurised Water Reactor (PWR). Both the AGRs and the PWRs operate on fuels with higher concentrations of uranium-235 and thus more reactive than natural uranium: the method of enrichment of the uranium-235 isotope at Capenhurst in Cheshire is outlined in Appendix 1.4. Typical compositions of these fuels are the dioxide of uranium, enriched to 2.2% and 3.3% uranium-235 respectively. The AGR fuel is canned in stainless steel; the can of the PWR can be the same, but more often is an alloy of zirconium which absorbs neutrons less readily than stainless steel. Cooling of the AGR fuel and graphite moderator is by carbon dioxide, as in the Magnox reactor, whereas the PWR is both cooled and moderated by high-pressure water, which yields its heat to a separate water/steam circuit driving the turbines. The steel in the AGR cans and the hydrogen in the PWR water both act as neutron absorbers and so are important factors determining the degree to which the fissile content in the fuel must be concentrated. A further type of reactor, the Steam Generating Heavy Water Reactor, which uses heavy water as the main moderator and ordinary water for cooling, was built at Winfrith in Dorset and has been operating successfully for 20 years. However, further development of a series of commercial stations was discontinued in favour of AGRs.

Spent fuels from AGRs and PWRs have uranium-235 concentrations close to that of natural uranium. It may still be economic therefore to recover this uranium from spent fuel for eventual re-enrichment at the Capenhurst plant. The quality of plutonium in the above spent fuels is slightly lower than that from Magnox reactors, i.e. it has a lower percentage of fissile isotopes plutonium-239 and 241.

1.2.4.3 <u>Fast Reactors</u>. In a Fast Reactor, fuel in the reactor core can have typically 20% plutonium and 80% uranium in the form of oxides, canned in stainless steel. This is cooled by liquid metallic sodium, which passes on its heat to a second circuit of sodium, which, in turn, passes heat to a steam-raising circuit. A considerable proportion of the neutrons escape from the core, but these can be put to use by surrounding the core with a 'breeder' of uranium oxide, typically material of below natural enrichment from Capenhurst. This absorbs neutrons to generate plutonium from uranium-238: though there is a reduction of plutonium content in the core during a fuel cycle, the net output of plutonium from the reactor as a whole can be greater than the input. The system, in fact, can not only be self-sustaining in fissile isotopes, but also used to provide fuel for the startup of other Fast Reactors. The stock of plutonium needed to sustain a Fast Reactor fuel cycle must provide sufficient plutonium not only for an initial charge but to keep the reactor going while its spent fuel goes through the complete reprocessing and fuel fabrication cycle (see, for example, the diagram of the Fast Reactor Equilibrium Cycle in Section 2.5). Such a stock of plutonium for each of the early Fast Reactors must come from the spent fuel of Thermal Reactors.

We shall look in more detail at the flows of the more important nuclides in the possible future nuclear cycles in Chapter Two.

1.3 COAL-FIRED POWER

The principles of obtaining power from coal are well known and simple. Coal is derived from organic matter containing varying amounts of minerals; this mixture was laid down from vegetation and for many thousands of years subjected to high temperatures and pressures. The end product is a mixture of carbon, hydrocarbons and inorganic matter, the relative proportions of which can vary widely from mine to mine. In the Selby (Yorks.) complex, for example, the coal produced has very little mineral impurity (or spoil) and requires no washing. On the other hand, the Belvoir (Notts.) mine has about one tonne of spoil per two tonnes of coal extracted. It has been estimated that there are sufficient coal deposits in the UK to supply the whole of the national power requirements for two or three centuries.

In current power station furnaces, combustion of coal leaves so fine an ash that it 'flies' with the combustion gases and hence is known as fly ash. Most of this ash is brought down by electrostatic precipitation. The remaining gaseous effluent is directed to tall stacks to avoid hazardous levels of chemicals in the local atmosphere. In future, limestone beds may be installed to trap acid oxides either during combustion or from the flue gases (Longhurst, 1987): the former could be done in a fluidised bed (Fluidised Bed Combustion or FBC), where sand and limestone particles would be subjected to a stream of air sufficiently fast to keep them mobile, but not enough to carry them away in the gas flow. Finely powdered coal, together with makeup sand and limestone, would be directed onto the bed, which has been initially heated up by burning gas. The intimate contact of air and coal would provide efficient combustion; sulphur impurities in the coal would be oxidised, but instead of being released as sulphur dioxide gas as in current coal-fired power stations, they would be trapped by the limestone as solid calcium sulphate. Several methods of Flue Gas Desulphurisation (FGD), using alkali to pick up acid oxides, are currently being assessed - the Wellman Lord and limestone-gypsum processes, and seawater scrubbing at coastal stations. Other processes than FBC and FGD are being investigated to control stack emissions (Barrett, 1986; British Coal, 1987), such as:

(a) delayed mixing of air and fuel, which reduces the formation of nitrogen oxides,

(b) flue gas denitrification with ammonia,

(c) Pressurised Fluidised Bed Combustion (PFBC), which can be used to drive a gas turbine before the steam turbine, and

(d) combined coal gasifiers with gas turbines.

FGD inevitably reduces the power station efficiency (by about 2%), due to the extra energy used to drive the stack gases through the traps. On the other hand, PFBC gives increased efficiency (about 1.3%) due to the power from the gas turbine (NCB, 1985).

At present, well over 75% of all the coal produced in the UK is consumed in power stations. This requires considerable transportation and, in practice, the power stations are often located near to coal sources: for example, the Selby mine and its associated power stations are on a 'merry-go-round' short distance circuit. Stations are then usually too far from the coast to use seawater; consequently, cooling water must be recycled and itself cooled by large air-cooling towers. Even so, the loss of water by evaporation and purge of accumulating impurities is considerable, so that proximity to a large river is desirable (see Fig. 9.1).

1.4 FUSION POWER, RENEWABLE SYSTEMS AND ENERGY CONSERVATION

1.4.1 General

Whereas coal-fired and nuclear power are established processes, either of which is capable of supplying the whole of the UK power requirements, there are several systems which either need further development, such as fusion power, or have features which limit the extent of their application. A basic constraint underlying some of these limitations is the impracticability of storing electricity on a large scale. Indirectly, this can be done by pumping water, at times when the electricity demand is low, up a mountain to a storage lake, from where it is released to generate electricity from turbines at peak demand periods. Examples are at Dinorwic in Wales and Loch Awe in Scotland: however, given the geography of the UK, it is hard to suggest other attractive locations for such pumped storage. The creation of hydroelectric schemes in addition to those currently in operation is similarly constrained. It follows that systems based on naturally renewable energy sources such as windmills, wave machines or tidal barrages, can only supply electricity intermittently, there being no practical large-scale method of smoothing out the natural cycles.

The above systems have been grouped together in this section because of their current limitations, but this does not discount the possibility that there may be some technical breakthrough or change of circumstances which could enhance the extent and timing of their application in the future. Each of them is outlined briefly below with their level of development or inherent limitations.

1.4.2 **Fusion Power**

In Section 1.2.1, the possibility of creating energy by fusing together the nuclei of atoms was raised. Research into the development of this form of power has been in progress now for several decades. The basic requirement is to create a 'plasma', i.e. to strip atoms of their electrons with the resulting nuclei moving at such very high speeds that collisions can bring about fusion. Such conditions, brought about in the plasma by passing large electric currents through it, are equivalent to a temperature in the region of 100 million °C and normal containment by structural materials is out of the question. Magnetic fields are the common approach to constrain the nuclei in a small enough volume that an adequate rate of fusion is achieved. An important safety feature of fusion plasma is that, if their containment fails, they touch the walls and collapse; the total heat emitted thereafter is only sufficient to raise the temperature of the surrounding material by a few degrees. Further, the only radioactive nuclide in the plasma is tritium, and estimates of its release are an order of magnitude lower than from the Sellafield site or the Canadian Pickering reactor (which uses deuterium in its moderator) (Hancox, 1987).

Largely because of the difficulties of achieving the necessary magnetic containment, present forecasts suggest that fusion power will not be commercially viable before the middle of the next century.

Currently the earliest promise of a source of fusion power stems from the reactions below, where tritium and deuterium are hydrogen isotopes of mass numbers 3 and 2 respectively, and lithium is an element with an atomic number of 3.

Tritium + Deuterium -----> Helium-4 + Neutron (F)

The tritium would be produced by the reaction:

Lithium-6 + Neutron -----> Helium-4 + Tritium (G)

Lithium-6 is present at about 7.5% in natural lithium in readily available ores. Initially the neutrons for reaction (G) could come from fission reactors; later on, tritium for reaction (F) could be produced by reaction (G) in a 'blanket' of lithium surrounding the gas 'plasma' in which the fusion reaction (F) creates both energy and neutrons. Energy from (F) and (G) and the slowing down of neutrons by lithium atoms can be passed from the blanket to a conventional steam cycle to raise electricity. There is effectively an unlimited source of deuterium in the hydrogen in water, even though its isotopic abundance is only 0.014%. Heavy water, which is deuterium oxide, has been produced commercially for several decades, mainly for the Canadian nuclear programme: in this process, natural water is enriched in deuterium isotope by a multi-stage process, involving hydrogen sulphide as an intermediary which is not consumed. It is easily estimated that a large fusion reactor would only require a few tonnes of fuel per year.

A more advanced form of fusion power would require deuterium only:

Deuterium + Deuterium ----> Tritium, Helium-3,
Helium-4, etc. (H)

these alternative primary reactions giving helium-3 plus a neutron or tritium plus a proton, then a secondary reaction of tritium with deuterium to give helium-4 plus a neutron.

Clearly, fuel supplies would be even more readily obtainable than for the tritium-deuterium reaction: moreover, there are less neutrons formed than in reaction (F), so there would be less radioactivity induced in surrounding structural materials. However, reaction (H) needs even higher plasma temperatures than (F), so the difficulties of commercial power development appear even greater.

With little transport requirements for lithium ores or heavy water, fusion reactors would have similar flexibility in siting to fission reactors, including construction on the coast to use sea-water in once-through condenser cooling.

1.4.3 Renewable Systems

By definition, renewable power sources require no fuel as their energy is replenished naturally. There is no constraint on their siting, therefore, because of any requirement to transport fuel. Rather, the location of the power unit is usually determined by natural features. A useful recent review has been published by the UK Department of Energy (DoEn, 1988b).

Hydroelectric power, derived from the potential energy of water in lakes high above nearby land, is, of course, a well-established process. Electricity is produced by directing water through turbines, whose shaft rotation drives generators as described in Section 1.1. In some countries, e.g. Norway, the mountainous terrain allows a considerable proportion of the country's power supply to be obtained hydroelectrically. In the UK, the mountainous regions of Scotland and Wales have been exploited in this manner. Overall, however, only a very small fraction of power is obtained in this way (0.1 GW(E) or about 0.2% in England and Wales) and it is difficult to find further sites capable of installing economic units.

In the last few decades, strenuous efforts have been made to develop power systems based on renewable cycles of wind and water. Unfortunately, these can inherently only yield a supply which fluctuates with the natural driving forces, so that a steady and reliable power supply cannot be derived from such sources (Elkington, 1984). Although it follows that the bulk of power supply to a highly industrialised country like the UK must stem from other types of system, which respond reliably and rapidly to demand, the renewables can be 'blended in' the UK Grid; it appears worthwhile to supply perhaps 10-20% of peak power requirements in this way (Milborrow, 1985). A consequence, however, is that other types of power plant have to be operated at part load ready to cut in when such variable sources as wind and wave power are unavailable; this incurs an inherent economic penalty.

The tidal sequences induced by the Moon have long been used as a form of hydroelectric power in a number of regions round the world. For a reasonably economic system, the range between high and low tides must be high - say about 10 metres. This often occurs naturally where the narrowing of an estuary causes the in-flowing sea to be 'pushed up' as the sides of the flow are constricted. Storing the water behind a barrage then allows outflow on the ebb tide to be used to generate electricity. One such scheme has been operating for many years at La Rance in Brittany. In the UK, the Severn Estuary is a sufficiently attractive site that desk studies have already been carried out to establish the feasibility and order of cost. Because of the cyclic supply, although a peak output of 7.2 GW(E) was possible, the Severn Barrage could replace about 1.1 GW(E) of 'steady' alternative supply, only 2% of the UK requirement. Sites at Morecambe Bay and the Mersey Estuary would only average 0.5 and 0.15 GW(E) respectively.

There are several methods proposed for abstracting power from the motion of waves on the sea, e.g. Salter's Duck, the SEA Clam, the Lancaster Flexible Bag and the NEL Oscillating Water Column (for further details, see Elkington, 1984). These involve various energy transfer arrangements, such as floats which move up and down on the waves or changes in air pressure transmitted to turbines as water depths oscillate with the waves. All the methods obviously require a coastal location, preferably selected for good wave characteristics but, as an unfortunate consequence, severe storm conditions can occur. Considerable construction may be necessary, with long structures to derive an acceptable level of power. Further, the units may be spread over a wide area, albeit of sea. For example, it is calculated that the 'clam bag' unit of 10 MW(E) would require a 275-metre spine with 10 clams attached; the equivalent of a

2 GW(E) power station would then be 55 kilometres long.

Wind power requires even larger areas, in order to avoid interaction between units. For example, the replacement of currently planned UK nuclear power by wind machines would need 10,000 square kilometres, 4% of the rural land area of England and Wales (Elkington, 1984). Though there is a possible market for replacing diesel power units in remote parts of the UK, the impact of general application on land would therefore appear unacceptable. Offshore siting, say at depths of 10 to 50 metres is therefore indicated, but this, in turn, introduces extra complexity in construction and maintenance, with associated increases in costs. Flood (1987) has suggested that the UK could construct 2,000 wind machines, each of 8MW(E) in shallow coastal waters and several thousand smaller units on land.

An inherent technical feature is the sensitivity of power output to wind speed. No power is produced with winds of less than about 15 mph, but output is proportional to the cube of the wind speed thereafter. Environmental problems include electromagnetic interference to radio and TV transmissions, noise, visual effects and the impact on bird life. Nevertheless, the UK Central Electricity Generating Board (CEGB) has been increasing its research commitment to wind power (Milborrow, 1985): one interesting discovery is that the annual averages of wind speed on the same site have been found to vary between years by as much as a factor of 2.

1.4.4 **Conservation of Energy**

The title of this section is meant to be interpreted in a broad sense, i.e. conservation includes any means by which power requirements are reduced. This, however, does not include methods such as pumped storage described in Section 1.4.1, which merely delay the time when power is used.

Two forms of power in use today in many countries are solar power and geothermal power. Natural conditions in the UK, however, suggest little promise of either of these supplying a significant fraction of UK power demand. The efficiency of transforming energy from photovoltaic cells is so low that the latter would have to be as cheap as ordinary pavement slabs to provide reasonable solar power costs (Cohen, 1983) even in the US, which has a far higher level of sunshine than the UK. The temperature of rocks, even in relatively favourable locations in Cornwall, requires drilling to 2 and 6 kilometres in order to heat circulated water to 80° and 200°C respectively (Batchelor, 1985). Such low temperatures would make it difficult for electricity to be produced efficiently. It is surprising to find, therefore, that anti-nuclear groups have suggested that thirty Hot Dry Rock Schemes, each of 12 MW(E) capacity, might be installed in the UK by about the turn of the century (Flood, 1987). This is in spite of the fact that the amount of useful heat abstractable round a borehole is finite, so that new holes must be drilled periodically. On the other hand, there is a small potential to replace electrical heating by solar heating in buildings, particularly in the south east of England. Commonly, panels with black surfaces would absorb solar radiation; water circulating through the panels would transfer the heat induced by the radiation to radiator systems within buildings for space heating. Similarly, warm water from geothermal sources could be useful for district heating in some locations.

When heat is created to raise steam, the latter can be used either to generate power or to distribute heat or both. In the latter case, when the demands for heat and power in an industrial area have broadly similar patterns, Combined Heat and Power (CHP) often provides a satisfactory and economic solution. In fact, over two-thirds of private electricity in UK industry is generated in this way: private power without use of the heat is rarely economic. On the other hand, normally only about a third of power station fuel energy is converted to electricity, the majority being lost in cooling water and stack gases. CHP can therefore more than double the efficiency of fuel usage because less electricity is needed by consumers for heating. However, the heat and power demands by consumers do not always follow a similar pattern; moreover, there is a commitment to maintaining district heating schemes once initiated, even if more attractive power processes are developed. Further, because of the cost of transmitting and insulating hot water over long distances, it is advantageous to site the power station near large centres of population. Nevertheless, investigations by the CEGB have revealed locations where CHP could be attractive.

Hartlepool and Heysham nuclear stations provide examples of these: studies are also continuing into CHP possibilities in a number of large cities (Dart, 1985).

A variant of CHP for the future could occur if large industrial concerns installed their own CHP schemes to meet mainly heating requirements; surplus electricity could then, in principle, be fed into the UK Grid.

Application of CHP has been widespread in other European countries and Scandinavia, partly through Government subsidies, with a view to reducing oil imports. One reason for the lower use in the UK is that natural gas for home heating is relatively cheap. There is also a bigger proportion of home-owners in the UK; CHP is less attractive for such consumers than for urban communities in large blocks of flats.

Finally, reductions in energy requirements may be achieved in the future by using higher efficiency equipment (kettles, cookers, etc.) and better room insulation. This has been found to provide useful per capita fuel consumption reductions in France and West Germany. An energy efficiency demonstration scheme was set up in 1978 by the UK Department of Energy to encourage new energy savings in industry and buildings (Currie, 1988). For each £1 of a Government grant, the object was to save £5 per year of energy. It is felt that a saving of 20% in total UK energy consumption per unit of Gross National Product by the year 2000 is realistic. Building services, paper and textile industries and laundries appear particularly promising areas. Much more drastic measures are suggested by 'Green' pressure groups. Friends of the Earth (FOE) recommend 'hefty grants' towards better insulation, incentives for increased rail travel in preference to building more motorways, etc. (Porritt, 1984).

1.5 **SUMMARY**

The Earth consists of elements from hydrogen (atomic number 1) to uranium (atomic number 92), many of which are residues of a supernova explosion billions of years ago. The lighter nuclei of these atoms can be fused together at very high temperatures, e.g. two nuclei of deuterium (mass 2) joining to give a nucleus of helium (mass 4), simultaneously releasing energy. Nuclear energy can also be produced by splitting heavy isotopes such as uranium-235 and plutonium-239 with neutrons. Chemical energy, bound up in coal originating from plants synthesised in sunlight and then compressed and heated for many thousands of years, can be transformed into heat by burning in air. The heat derived from these processes can be converted to electric power by boiling water to drive steam turbines. The mechanical energy available from the waves, wind and the tides, requires no intermediate heat conversion step to turn it into electricity. However, their dependence on the natural behaviour of the winds and the Moon means that power production by these means is necessarily intermittent and there are no practical ways of storing electricity to smooth out such uneven production. Nuclear fission can be operated to produce power in various cycles, from simple burnup, mainly of uranium-235, in Thermal Reactors, to the burnup of plutonium recycled repeatedly through Fast Reactors. Recovery of plutonium by chemical reprocessing of spent fuels is necessary for Fast Reactors: enrichment of uranium in the 235 isotope is necessary for most types of Thermal Reactor fuels. Nuclear fusion is based on simple reactions but the containment of gases for long enough periods at the very high temperatures required for power production is difficult: commercial development therefore seems many decades away. Basic fuels for fusion, deuterium and lithium, are abundant, and recycling and breeding intermediate tritium appears straightforward.

Nuclear fusion and fission power plants need little transport for the small masses of fuel required, so they can be conveniently sited on the coast with direct cooling for their steam condensers from seawater. Coal-fired power stations, on the other hand, require the transport of very large quantities of fuel and therefore tend to be near coalfields, with air-cooling towers for steam condensation and with makeup water from rivers.

Tidal and wave power schemes must of necessity be on the coast or at sea: wind systems, because of their large area requirement, may well be sited offshore. Geothermal plants will probably not be economic for power production but may supply hot water for community heating; this is unlikely to be economic over long transmission distances and so the schemes will be confined to local 'hot dry rock' regions. Solar heating would probably not involve distribution, but would be constrained to direct production and use in buildings in the south of England where there are more hours of sunshine. Conservation of energy, e.g. by increased insulation, can also reduce demands of electricity for heating purposes.

Appendix 1.1

THE EVOLUTION AND DISTRIBUTION OF CHEMICAL ELEMENTS

A.1.1.1 **The Formation of the Earth**

When a gas cloud in a galaxy collapses to a core of much smaller volume under its own gravity, heat is emitted. If the mass of gas in the core is less than about one-tenth of the mass of the Sun, it is too small to cause nuclear fusion of its hydrogen to helium. The temperature cannot then rise high enough for the object to shine brightly; this type of star, of which the planet Jupiter is an example, is therefore known as a 'brown dwarf'. Our Sun, however, has sufficient mass to 'switch on' nuclear fusion and so radiate light; at an age of about 5 billion years, it has now converted about half of its original hydrogen to helium.

If the core mass is about ten times that of our Sun, fusion can proceed beyond helium (mass number 4) as far as iron (mass number about 56). Eventually, as the light elements are consumed, resistance to gravity is suddenly overcome and there is a very rapid contraction to a superdense core. Heavy elements (some with mass number over 200) are formed in the subsequent explosion and reaction with the outer gases; this phase is seen in the sky as a supernova, brighter than a billion Suns, but fading with a half-life of 78 days, corresponding to the main source of the radiation, cobalt-56. The core eventually contains only neutrons and is therefore commonly known as a 'neutron star'. The heavy elements are dispersed in space; some of these, from a supernova long ago, were trapped in the Sun's gravitational field and collected up with primordial gases in the formation of our Earth.

Larger initial masses of thirty to fifty times that of the Sun have an even shorter life, creating such an enormous gravity that nothing can escape, not even light waves, so causing the 'black holes' of space.

A.1.1.2 **Nuclide Changes in the Biosphere**

The section of the Earth which interacts with man is known as the biosphere - the atmosphere, the oceans, the rivers and inland lakes, and the surface of the land. Clearly, in the colossal neutron reactions of a supernova, not only our naturally known elements were formed, but also the products of their reactions with neutrons. However, if the nuclei formed were unstable, they would decay away. Such decay is at a steady fractional rate, with a specific time for the nuclide known as the half-life, within which half of the nuclei will change. For example, the element potassium, atomic number 19, has a naturally occurring isotope of mass number 40, potassium-40; this has a half-life of 1.3 billion years. The Earth has existed for approximately 5 billion years or about 4 half-lives of potassium-40, so that only about one-half to the power 4 or one-sixteenth of its nuclei would survive this length of time. A nucleus of plutonium-239, on the other hand, has a much shorter half-life, 24,390 years. A similar period of 5 billion years is in this case equivalent to about 205,000 half-lives; the chance of a plutonium-239 nucleus surviving this time is therefore only about 1 in 10 to the power 90. In fact, only 9 radioactive isotopes of atomic number less than 80 occur in significant concentrations naturally, with half-lives varying from 1.3 billion years for potassium-40 to 500,000 billion years for vanadium-50.

The above loss of an unstable nuclide by radioactive decay may be counteracted in nature by the creation of more of its nuclei in three ways. In the first of these, the local concentration of fissile nuclides may become sufficiently high as to produce a significant flow of neutrons (an initial source of these can be the spontaneous fission of some nuclides); these can then react to create heavier isotopes. An example of this occurred at Oklo in Central Africa many millions of years ago, when uranium-235 fissioned, creating neutrons, some of which reacted with uranium-238 to form plutonium-239. The latter and fission products survive in detectable concentrations at Oklo even today. It is quite probable that in the early stages of the Earth's existence, before uranium-235 decayed down to its much lower abundance today, there were other localities where natural fission reactors created both heavier isotopes and fission products. Secondly, in very dilute concentrations, about 1 atom for every 100 billion uranium atoms, plutonium is present in most uranium ores; here, alpha particles react with light nuclides to give neutrons which convert uranium-238 to plutonium-239 via the 'conventional' route occurring in a nuclear reactor (Katz, 1986). The third method of creation of radioactive nuclei on Earth is by the decay of 'parent' nuclides. A 'chain' then develops with a sequence of unstable nuclides. A particularly important chain, starting with uranium-238, is described in Appendix 1.2 and illustrated in Fig. 1.1, with an associated list of half-lives in Table 1.2. It should be observed that none of the half-lives, except that of uranium-238, is long enough to allow a significant fraction of a nuclide to persist over a period of 5 billion years, i.e. the Earth's age. All of the nuclides shown in Fig. 1.1 therefore occur in uranium ores because of the survival of uranium-238 and its continuous production of the chain. Over many millions of years they have all reached equilibrium in the chain, i.e. they have the same rate of decay as the parent nuclide uranium-238.

The half-life of nuclides and their neutron reaction characteristics are important parameters which in most cases are accurately known; consequently, good estimates by relatively simple calculations can be made of nuclide concentrations in ores and in fuels, both inside reactors and in process plants.

A.1.1.3 The Physical Distribution of the Earth's Elements

As the earth cooled down, a surface crust was formed and many elements were chemically bound up therein in minerals. (Some of the radioactive nuclides naturally occurring in rocks are discussed in Appendix 1.2.) Later, condensation of water began the hydrological cycle of evaporation into the atmosphere, precipitation of rain onto the land masses and flow back to the oceans. Runoff of the precipitation into streams and rivers, together with other erosion processes, removed enormous quantities of solids from the land surface. Further, percolation through the ground leached out soluble salts. The solids and salts eventually reached the sea, establishing the sea-bed sediments and salinity of the seas familiar today. Some sediments of ancient lakes and seas have been lifted up through crustal movements, forming sedimentary and metamorphic strata as found today. Coal deposits and oil-bearing strata are among examples of these; other natural processes, in addition to the deposition of toxic non-radioactive minerals, have led to the incorporation of 800 million tonnes of uranium and 300 tonnes of associated radium in the top 500 metres of British rocks, a factor discussed further in Appendix 4.4. Estimates of the levels of some of the more important radionuclides in the oceans are set out in Appendix 1.2. In one sense, therefore, it is difficult to justify the expression 'pollution of the sea', since the latter is effectively rainwater polluted by countless natural cycles of erosion and leaching. Assessment of the effects of man-made effluents should therefore be confined to whether there has been a significant increase in hazard to animal life, especially man.

The geographical evolution of the UK has some importance in later discussions of the disposal of nuclear wastes. It is only a few thousand years since Britain and Ireland became separated from the rest of Europe by the North Sea and English Channel. Erosion on Britain's east coast will continue in the future, so that some land areas there will become part of the sea-bed. On the west coast of Britain, on the other hand, the Irish Sea will recede sufficiently in the next few thousand years for Britain and Ireland to be joined by land. Sudden geological changes through earthquakes or volcanoes are of extreme low probability at any given place in the UK over the same timescale.

Appendix 1.2

THE NATURAL DECAY SCHEME FOR URANIUM-238

The decay of uranium-238 is important in determining the level of radioactive impurities in uranium ore; it is a good example of the characteristics of radioactive decay sequences which occur naturally in other chains from thorium-232 and uranium-235 and also from fission products and heavy nuclides after they are removed from neutron irradiation in reactors.

Fig. 1.1 shows the uranium-238 decay chain with atomic number plotted against numbers of neutrons in the nucleus: corresponding nuclide characteristics are listed in Table 1.1. Isotopes of each element therefore lie on the same horizontal line: nuclides of the same mass are on an upward left sloping diagonal, often connected by beta decay (no change in mass and a unit increase in atomic number). Alpha decay (a fall of two units in both atomic number and neutron number) results in a downward left-sloping diagonal.

The uranium isotopes 238 and 234 decay into first the daughter thorium isotopes 234 and 230, then granddaughters protoactinium-234 and radium-226. The latter is slow to reach equilibrium, i.e. 'match' its rate of formation with decay, since it has a half-life of 1,600 years. Its own descendants down to polonium-214 have short half-lives and so soon approach the same activity as radium-226: there is then a slight 'holdup' at lead-210 (half-life 22.3 years) before the final stages of decay to the stable isotope lead-206.

A common rock in the UK, granite, can contain typically 5 parts per million by weight of uranium, so that a cubic kilometre of granite could hold about 13,000 tonnes of uranium-238, emitting 160 TBqs (4,300 Ci) of radioactivity. This will, of course, be in equilibrium with its 13 products of decay as shown in Fig. 1.1, so that each of these will be emitting radioactivity at the same rate, giving a total for the chain of 13 x 1.60 = 2,000 TBqs (54,000 Ci). These figures, and those below for the ocean, are useful natural yardsticks against which to compare radioactive levels of nuclear wastes.

Seymour (1971) gives data on the total volume of the oceans (1,700 million cubic kilometres), the concentrations of radionuclides in solution therein and also in the sediments of the sea-bed. Not all of the uranium-238 decay chain of Fig. 1.1 stays in solution in the sea; most of the long-lived thorium-230 is precipitated into the sediments. The rest of the chain, however, is in solution with the uranium-238.

The total mass of the latter is 5,000 million tonnes(!) emitting 60 million TBqs (1,600 million Ci) of radioactivity. Using the data from Seymour for the other nuclides in the uranium-238 chain, the total activity in the oceans from this chain is about 300 million TBqs (8,000 million Ci) (about 180 Bqs per cubic metre or 5 Ci per cubic kilometre), of which 6 million TBqs (160 million Ci) come from radium-226. Other chains from thorium-232 and uranium-235 have much lower total activities. These figures will be used in later comparisons of the radioactive arisings from nuclear fuel cycles.

Appendix 1.3

MAIN NUCLEAR REACTIONS IN POWER REACTORS

Fig. 1.2, with the characteristics of nuclides as listed in Table 1.2, shows the more important types of reaction occurring in power reactors. Only those isotopes readily fissionable in Thermal Reactors, in particular uranium-235, plutonium-239 and 241, have been shown as fissioning, though, of course, in a Fast Reactor, all the heavy nuclides fission to some degree. Alpha and beta decay follows similar diagonal directions to those shown in Fig. 1.1: neutron capture does not, of course, change the element and so results in a horizontal step to the right. In addition, there is a further type of reaction, where a neutron reacts to produce two neutrons from a nucleus causing left horizontal changes for uranium-238 and plutonium-239.

Some nuclides are formed by more than one successive neutron reaction, so their concentration increases more than linearly with the time of irradiation. On the other hand, once reactor fuel is discharged there are virtually no further neutron reactions and the shorter-lived nuclei decay away, e.g. uranium-237, neptunium-238 and neptunium-239, are effectively eliminated during the normal 'cooling' period of several months out-of-reactor storage before fuel reprocessing. Nuclides of medium length half-lives, such as curium-244 with an alpha half-life of 18 years, will not usually be important in the disposal of wastes since barriers to their movement are easily designed to last several centuries.

Appendix 1.4

THE CREATION AND CONTROL OF NUCLEAR REACTIVITY

A.1.4.1 Enrichment of Fissile Isotopes

Section 1.2.2 introduces some basic reactions in nuclear power reactors; effectively deriving continuous power from these depends on neutrons being produced at the same rate as they are consumed. Clearly, the higher the concentration of fissile nuclides and the lower the level of neutron-capturing components, the easier it is to sustain a steady neutron reaction rate and corresponding power output. Of the two major uranium isotopes in nature, uranium-235 at 0.72% abundance is fissile, whereas uranium-238 at 99.28% abundance is effectively non-fissile in Thermal Reactors and only slightly fissile in Fast Reactors.

In order to operate Advanced Thermal Reactors (see Section 1.2.3), the fuel must be enriched in the uranium-235 isotope. The principal method of uranium enrichment used at Capenhurst near Chester in the early nuclear years was to prepare uranium hexafluoride, or 'Hex', which is a gas at slightly above ambient temperature, and cause it to diffuse through porous membranes. The rate of diffusion of the hexafluoride molecules is slightly faster with uranium atoms of mass 235 than for mass 238. The lighter fraction can then be passed on in an opposite direction to the heavier fraction at each stage of a large number of membrane cells. Hex of the required enrichment is then chemically converted at Springfields near Preston to the uranium dioxide pellets used in the fuel rods for the AGR. Hex of less than about 0.4% uranium-235 isotopic content is rejected from the plant to be converted into oxide for storage and possible use in Fast Reactor fuels.

Over recent years, the very expensive (in energy input terms) diffusion process has been replaced by a system of uranium enrichment called the centrifuge process. This involves a multiplicity of centrifuges which operate to concentrate the heavier uranium-238 component in Hex towards the walls of the centrifuge. At each centrifuge stage the separation of the isotopes of uranium is fractionally very small and so many stages are necessary to manufacture a product of the required enrichment for AGRs and PWRs.

A.1.4.2 **Rapid Control of Fuel Reactivity**

Since the average life of neutrons in a reactor is short, their concentrations can vary very quickly, giving unacceptable surges of power. Fortunately, a small but significant fraction of the neutrons arise from the decay of a few fission products: these have short half-lives, about a minute, but this is ample time for neutron-absorbing control rods to adjust their position to maintain a steady level of power. In the accident at Chernobyl, the reactor operated in a condition where the fission or 'prompt' neutrons multiplied faster than the control system could respond and overwhelmed it for short periods, giving surges of energy similar to those in non-nuclear chemical explosions. Further chemical reactions then took place, with additional explosions (Gittus, 1987). This is a very different state of affairs from when the intent is to cause an atomic explosion, where the nuclear components have to be fired together at such great speeds that they all coalesce within a millionth of a second. In effect, Chernobyl, though creating a very serious spread of radioactivity, was a convincing demonstration that, even under the worst circumstances, repeated at several stages of the accident, a reactor cannot explode like an atomic bomb.

A.1.4.3 **Long-Term Changes in Fuel Reactivity**

Under normal reactor operating conditions, the change in reactivity (the ability of the fuel to react with neutrons) is slow. It is therefore easy to achieve control by gradual withdrawal of some of the neutron-absorbing control rods as the fissile isotopes are 'burnt up' and neutron absorbers in the fuel, such as many of the fission products, build up. Eventually, however, the reactivity of the fuel to neutrons becomes so low that it is desirable to replace it.

Since each fission yields a fairly precise amount of heat, the degree of fissioning of a fuel, and hence its reduction of reactivity, can be related to its overall heat output, usually expressed in terms of megawatt-days (MWD)/tonne.

If, for example, for each tonne (1,000 kilograms) of a section of fuel, there was an output of 10 megawatts of heat (10,000 kilowatts) for 100 days, the fuel irradiation would be 1,000 MWD/tonne. According to reactor type, fuel spends the equivalent of one to five years at full reactor power before it is spent. For atom bomb production in 'military' Magnox reactors, it is desirable for plutonium to be of 'high quality', i.e. with only a small percentage of the isotope of mass 240, produced by the secondary reaction (D) in Section 1.2.3. Consequently, buildup of plutonium-240 must be avoided by discharging the fuel at a relatively low irradiation - about 300 MWD/tonne. In a Magnox reactor, on the other hand, the same type of reactor, run for power purposes, now 'burns' fuel up to 5,000 MWD/tonne, a considerable improvement on its early performance a few decades ago, of only 3,000 MWD/tonne. The low quality, i.e. relatively high concentration of the mass 240 isotope in this plutonium is referred to as 'civil', denoting its suitability for burning in power reactors and not for making bombs. The plutonium in spent fuel from AGRs (18,000 MWD/tonne) and PWRs (33,000 MWD/tonne) is similarly only of 'civil' quality.

Table 1.1

Radioactive Characteristics of the Naturally Occurring Uranium-238 Decay Chain in Figure 1.1

Nuclide	Symbol	Decay	Half-Life
Lead-206	Pb206	Stable	Infinite
Lead-210	Pb210	Beta	22 years
Lead-214	Pb214	Beta	27 minutes
Bismuth-210	Bi210	Beta	5 days
Bismuth-214	Bi214	Beta	20 minutes
Polonium-210	Po210	Alpha	138 days
Polonium-214	Po214	Alpha	<1 second
Polonium-218	Po218	Alpha	3 minutes
Radon-222	Ra222	Alpha	3.8 days
Radium-226	Ra226	Alpha	1,600 years
Thorium-230	Th230	Alpha	80,000 years
Thorium-234	Th234	Beta	24 days
Protoactinium-234	Pa234	Beta	7 hours
Uranium-234	U234	Alpha	250,000 years
Uranium-238	U238	Alpha	4,500 million years

Table 1.2

Main Characteristics of Nuclides in Nuclear Reactors Illustrated in Figure 1.2

Nuclide	Symbol	Decay	Half-Life
Uranium-235	U235	Alpha	700 million years
Uranium-236	U236	Alpha	23 million years
Uranium-237	U237	Beta	7 days
Uranium-238	U238	Alpha	4,500 million years
Uranium-239	U239	Beta	24 minutes
Neptunium-237	Np237	Alpha	2.1 million years
Neptunium-238	Np238	Beta	2 days
Neptunium-239	Np239	Beta	2 days
Plutonium-238	Pu238	Alpha	88 years
Plutonium-239	Pu239	Alpha	24,400 years
Plutonium-240	Pu240	Alpha	6,540 years
Plutonium-241	Pu241	Beta	15 years
Plutonium-242	Pu242	Alpha	387,000 years
Americium-241	Am241	Alpha	433 years
Americium-242	Am242	Beta	152 years
Americium-243	Am243	Alpha	7,400 years
Americium-244	Am244	Beta	10 hours
Curium-242	Cm242	Alpha	163 days
Curium-243	Cm243	Alpha	30 years
Curium-244	Cm244	Alpha	18 years

Figure 1.1: URANIUM-238 NATURAL DECAY CHAIN

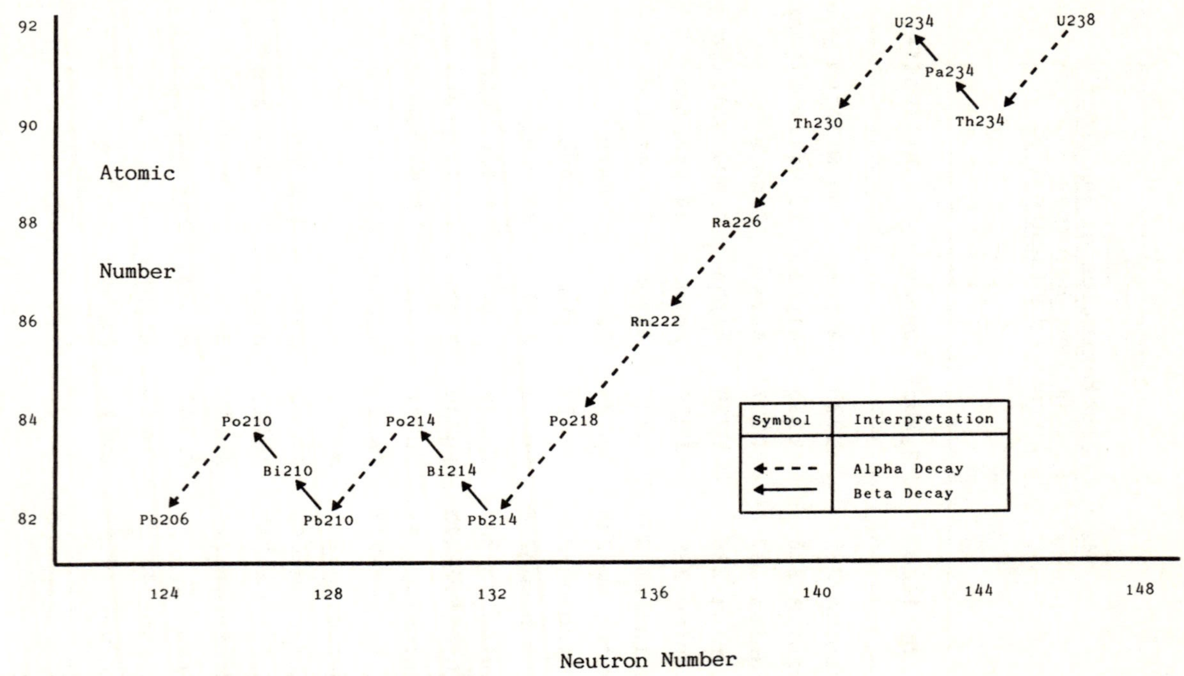

Figure 1.2: MAIN NUCLEAR POWER REACTIONS

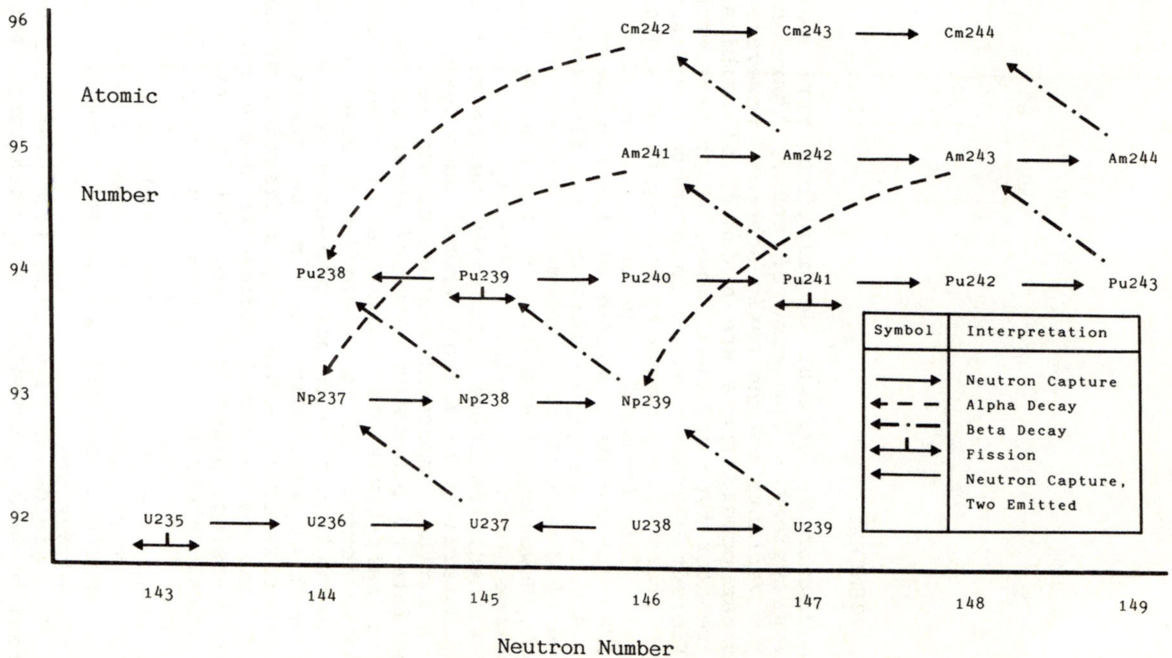

...DE FLOWS IN NUCLEAR FUEL CYCLES

2.1 GENERAL

In Chapter 1, general descriptions were made of power systems. For most of these, there are no major variants on the basic scheme. However, with nuclear power, there are a number of combinations of reactor type and fuel flows, which have markedly different characteristics in uranium requirements, fuel storage, reprocessing and waste arisings. If uranium-235 (natural abundance in uranium, only 0.72%) is the main source of fission in the fuel cycle, the usage efficiency in fissioning the natural isotopes of uranium will clearly be very low. However, if uranium-238, the dominant natural isotope, can be either directly or indirectly fissioned, the usage efficiency can be much higher and natural uranium requirements much lower.

A further important feature of any fuel cycle is the fate of plutonium, some of whose isotopes have long half-lives of thousands of years and are highly toxic. If it cannot be disposed of safely by convenient non-nuclear processes such as burial, it must at some point in time be 'burnt up' in a nuclear reactor: since only a fraction of nuclear fuel can be consumed before it needs renewal, a reprocessing plant is required to recover the residual plutonium to make into more fuel for further 'burning'. On the other hand, if disposal of spent fuel by non-nuclear methods is proposed, plutonium nuclides and their decay products become an important feature in safety aspects of the disposal system.

The yields of nuclides from the fission of uranium and plutonium are similar; for all practical purposes, therefore, each cycle to be described produces the same very high level of fission product activity (which, apart from a few long-lived exceptions such as technetium-99 and iodine-129, effectively dies away after one or two hundred years, plus the levels of long-lived isotopes of plutonium and americium to be listed below. Since many different safety analyses have shown that the critical property of a waste nuclide is its toxicity on ingestion, the changes after disposal are discussed in terms of this type of toxicity, not mass or radioactivity.

Below, the more important variants of nuclear fuel cycles are presented in turn, with discussions on their efficiency of uranium usage, the flows of plutonium and other key nuclides, and their potential toxicity on disposal. A final section compares the above features in the various fuel cycles.

For an excellent more detailed account of reactor characteristics and corresponding nuclide flows, see Farmer (1983) and Jones (1987), from which some figures have been derived for Table 2.1 at the end of the chapter, the remainder being calculated by the NTC computer routine 'ISOCYC'.

2.2 THE ONCE-THROUGH THERMAL REACTOR CYCLE

The basic fuel flows for this cycle are shown below.

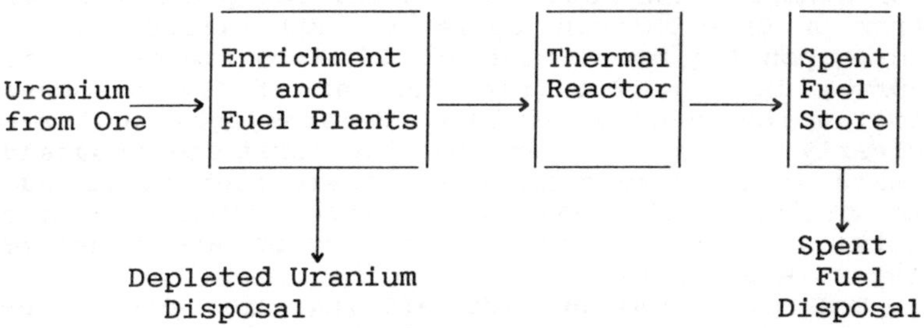

This system is similar to that of coal-fired power in that all discharges during fuel preparation and power station operation are treated as wastes: the only reason here for storage of spent fuel is to decay away much of its activity and so simplify the eventual disposal operations. The other waste flow, depleted uranium, is of very low activity: though a small fraction might be used in industry, much of it may be a waste product. Safe disposal of spent fuel depends on the long-term behaviour of the long-lived nuclides of plutonium and americium: the rate of arisings of these is given in Table 2.1, a short time after discharge from a PWR. A century or so after discharge, most of the plutonium-241 (half-life 15 years) will have decayed away to americium-241 (see Fig. 1.2). The latter then decays more slowly (its half-life is 430 years) to the very long-lived neptunium-237 (half-life 2.1 million years). Both americium-241 and neptunium-237 can therefore 'grow' before decaying after disposal. However, the levels reached by the latter never pose a significant hazard (as further discussed in Appendix 6.2) and so it is not listed.

One guide to the relative potential hazard of waste nuclides, the reasoning for which is detailed in later chapters, is to evaluate the results of dispersing their annual production in drinking water and noting the corresponding dose as if all the water was then humanly ingested. (A convenient unit here is the Annual Limit of Intake (ALI) per person or its multiple, one billion ALIs: the billion used hereafter is the US billion or one thousand million.) This evaluated dose is given in brackets in Table 2.1 after the kilogram arising for each isotope per GW(E)yr (a unit of power supplied, equivalent to 1 gigawatt or 1,000 megawatts of electricity delivered for one year): for example, the hazard potential for plutonium-240 from a Once-Through Cycle is 200 billion ALIs, corresponding to a mass of 74 kilogrammes. The curves in Fig. 2.1 (see the end of the chapter) follow the changes of these hazards against time: clearly, the main hazard for the first few thousand years is from americium-241, thereafter it is due to plutonium-239 and 240. The results for the total hazards of the various cycles are compared later in Section 2.7.

The plutonium nuclide arisings for the Once-Through will be seen in Table 2.1 to be high relative to other cycles. This is, of course, because the total plutonium in spent fuel is a waste product. Even so, it is worth noting that, though plutonium is not present in fuel as charged to the reactor, over 60% of that created is used up in fission before fuel discharge. To a considerable extent, therefore, the use of plutonium for power production already happens in the uranium fuels of Thermal Reactors and so the 'plutonium economy', that doom-laden phrase of the 'Greens' and the media, is with us now!

The maximum usage efficiency can be seen from Table 2.1 to be very low, at only 0.5% (Farmer, 1983): as a consequence, a 1 GW(E) PWR, for example, could use 3,500 tonnes of uranium in its lifetime.

2.3 THERMAL RECYCLE

The above title has been used, e.g. in Farmer (1983), to denote a system in which plutonium and uranium in spent fuels from Thermal Reactors are recovered and recycled through the same type of reactor.

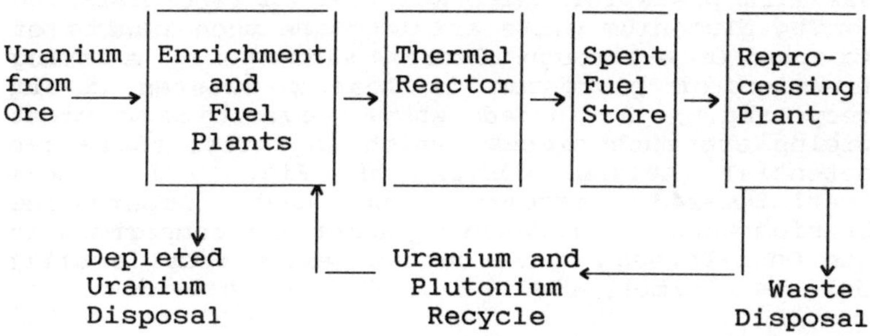

The figures for this cycle presented in Table 2.1 are for a PWR reactor. Separate fuel elements are assumed to be prepared containing either

(a) enriched uranium in the form of oxide as in the Once-Through Cycle, or

(b) mixed oxides (MOX) of plutonium and uranium, the latter either natural or depleted.

When the cycle is in a balanced mode, Farmer (1983) suggests that about a quarter of the fuel elements are MOX, containing about 8% plutonium of low quality (only 40% plutonium-239). Since this part of the fuel is already 'heavier' than uranium-238 in the Once-Through fuel, there are less steps needed to produce the isotopes americium-241 and americium-243; the yield of these is therefore relatively high. Some recycling of plutonium through Thermal Reactors has already been carried out in Belgium and French PWRs; only a few batches in small-scale experiments, however, have been done in UK AGRs. There may be a limit on the fraction of plutonium that can be accommodated in Thermal Reactors, because of different effects of uranium-235 and plutonium on localised control of reactivity, especially in AGRs. Large-scale 'burning' of plutonium in Thermal Reactors may thus need comprehensive demonstrations and maybe changes of reactor design before becoming a reliable process. The French utility Electricité de France (EDF) is committed to the use of MOX in PWRs; recently, a new Franco-Belgium Company, Commox, has been established to develop MOX fuels containing 3% to 5% plutonium. Trials of the fuel have already begun and, if successful, a manufacturing plant will be built at Marcoule, by 1992. Japan, too, has firm plans for using MOX fuel in reactors.

The plutonium waste arisings are much lower than in the Once-Through Cycle, since only a small fraction of plutonium is lost to wastes during recycling. As noted above, americium isotope yields are much higher, which is reflected in the potential hazard curves of Fig. 2.2, where americium-243 assumes increased importance. Uranium usage is reduced by about 30% compared with the Once-Through Cycle but the efficiency is still only 1% (Farmer, 1983).

2.4 THE FAST REACTOR LAUNCH CYCLE

The common fuel for Fast Reactors is MOX with plutonium at about 15-20% of the total plutonium plus uranium atoms. Although all heavy nuclides are more fissile in a Fast Reactor than in a Thermal Reactor, the nuclides plutonium-239 and plutonium-241 are still much more fissile in Fast Reactors than plutonium-240 and plutonium-242. For example, in a Fast Reactor plutonium-240 has only about 20% of the reactivity worth of plutonium-239. It has therefore become conventional to 'weight' the nuclides to derive an equivalent level of plutonium-239 (PuE) for each nuclide; the sum of these equivalents provides the same reactivity for input of new fuel as listed in Table 2.1.

A Fast Reactor system, initiated with (say) PWR plutonium, eventually produces more plutonium from its core plus breeder than is fed to the core. The early stages before this plutonium is recovered from Fast Reactor spent fuels can extend over a considerable period, so that it has been given a separate title - the 'Launch' Cycle (Marshall, 1980).

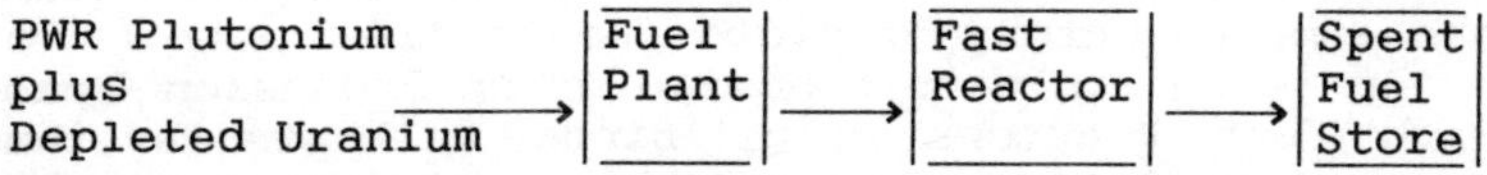

Ranging calculations show that waste arisings are intermediate between those of the Thermal Recycle and the 'Equilibrium Cycle' described in the next section. The system is, however, only a temporary phase: the spent fuel will eventually be processed, so wastes are not listed in Table 2.1.

2.5 THE FAST REACTOR EQUILIBRIUM CYCLE

When sufficient plutonium is to hand, the option is available to replace some of the breeder elements in a Fast Reactor with 'blanks', say of steel, so that no net breeding occurs. This 'Equilibrium' Cycle, shown below, therefore requires only an input of depleted uranium, which is mostly eventually converted to plutonium by neutron capture, thus acting indirectly as a fuel.

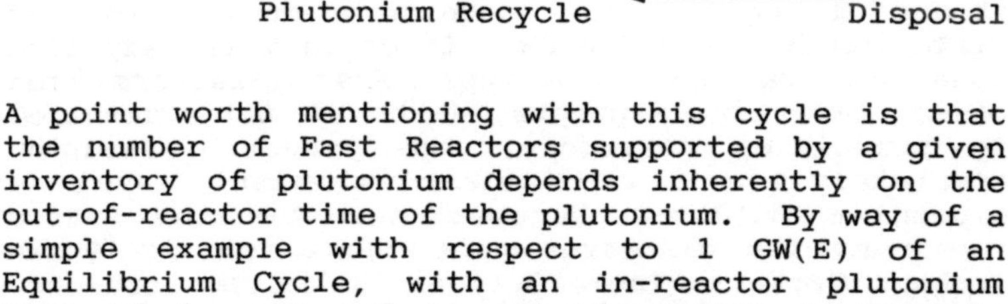

A point worth mentioning with this cycle is that the number of Fast Reactors supported by a given inventory of plutonium depends inherently on the out-of-reactor time of the plutonium. By way of a simple example with respect to 1 GW(E) of an Equilibrium Cycle, with an in-reactor plutonium time of 1 year and an out-of-reactor fuel cycle period of 5 years, Table 2.1 shows that a total of 6 x 1,900 kilograms PuE would be required, i.e. the plutonium nuclides present must be equivalent in reactivity in a Fast Reactor to 6 x 1,900 kilograms of plutonium-239. With an out-of-reactor period reduced to 1 year, however, only 2 x 1,900 kilograms of PuE would be necessary. In other words, the shorter fuel cycle time would allow three times as many reactors to be supported for the same inventory of plutonium as PuE.

It is possible to conceive of an Expansion Cycle with Fast Reactors only, since full use of the breeder will produce a surplus of plutonium which could be used to start up more reactors. The period necessary to yield enough plutonium for a total of twice as many reactors is known as the 'doubling time'. Typically, for an oxide reactor this could be 30 years. Nuclide production for this Expansion Cycle is not tabulated, since it would be similar to, but slightly lower than, that of the Equilibrium Cycle.

It can be seen from Table 2.1 that the Equilibrium Cycle is very efficient in uranium usage, estimated to be between 50% and 80% (Farmer, 1983). The production of very heavy nuclides is low: in particular, the hazard from americium-243 is never significant (Fig. 2.3). After the decay of americium-241, plutonium-239 and 240 provide the dominant hazard.

2.6 A THERMAL/FAST REACTOR CONTRACTION CYCLE

The usage of plutonium has been shown to be, in principle, feasible by Thermal Recycle. If, however, nuclear systems as a whole were to be run down, continued recycling of plutonium through Thermal Reactors with lowering quality and increasing concentrations relative to uranium could be difficult. On the other hand, since the core and breeder fuels of a Fast Reactor can be processed separately, adequate demonstration of plutonium incineration has effectively already been done in various prototype Fast Reactors and associated recycle plants in the UK and elsewhere. In principle, therefore, the breeder could be completely loaded with 'blank' elements; surplus plutonium could then be used up in the core. From the plutonium discharge rate shown for the Once-Through Cycle in Table 2.1, it is fairly straightforward to deduce that if this were reprocessed, about 250 kgs PuE would be available for a Fast Reactor. This figure is a close enough match to the net incineration rate in a Fast Reactor core of about 230 kgs (Farmer, 1983) to pair up Thermal and Fast Reactors of equal power in the scheme below.

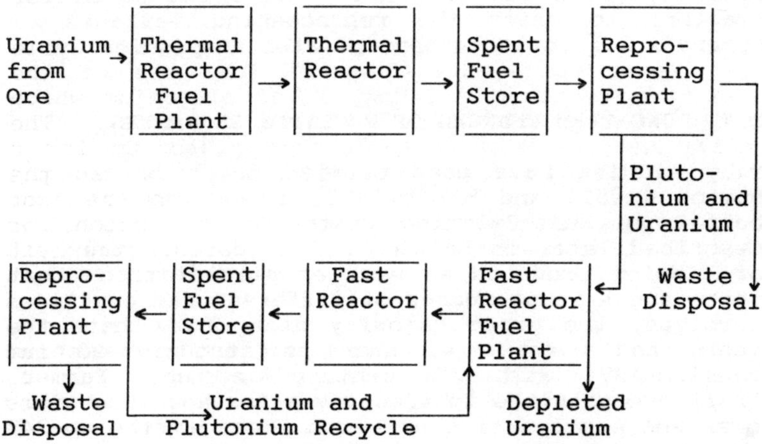

This cycle, being a rundown phase, does not achieve a steady balance in nuclides, so that detailed figures are not very meaningful. In broad terms, however, such arisings are intermediate between those of Thermal Recycle and the Equilibrium Cycle of the Fast Reactor.

Clearly, the timing of the production of plutonium is not very important, so that the system could be used either to control the buildup of plutonium or to run down old stocks. For example, if it were decided to run down UK stocks by 10 tonnes, about 40 GW(E)yrs of Fast Reactor operation would be necessary. It should be noted that unless new technology is developed, uranium must accompany plutonium in the core fuel, thus creating plutonium, albeit at a lower rate than the input of plutonium. Further, reprocessing is necessary at regular intervals: even with some 'inert' oxide replacing uranium oxide in the core, reprocessing would still be unavoidable and probably much more frequent to maintain a steady reactivity over the core. At the same time, the quality of plutonium would fall and a higher concentration of plutonium would be needed, which eventually could lead to problems in dissolving spent fuel in nitric acid during reprocessing.

There are obviously a whole range of Thermal-Fast Reactor combinations, from the 1:1 relationship above to higher ratios, such as 5:1, where some recycling in Thermal Reactors could be done before passing on low-quality plutonium to a Fast Reactor, to which the reprocessing variants of Appendix 2.2 could be appropriate.

2.7 LONG-TERM CHANGES OF NUCLIDES IN WASTES

Many studies have been carried out, for example Burton (1981) and Hill (1978, 1981), on the land burial of Heat-Emitting Wastes (HEW), which, as described later in Section 3.2.1, contain the bulk of fission products as well as americium. Though there are a large number of different radionuclides involved, the great majority decay away in a few years and even more, such as strontium-90 and caesium-137, within a century or two. It is therefore possible to identify a few key long-lived ones whose properties are such that, if the design can be shown to be safe for them, the rest are automatically safely contained. This is the reason for the emphasis in preceding sections, on the behaviour of the long-lived nuclides of plutonium and americium.

As mentioned earlier, each of the systems of Table 2.1 has been evaluated for the level of potential ingestion hazard from the listed isotopes. The curves of Figs 2.1 to 2.4 illustrate the significant contributors to hazard on a timescale from 300 to 300,000 years after their discharge from a reactor. It is important to note that the scale in all figures is, for convenience, logarithmic. In Fig 2.1, for example, the potential hazard at 300 years from americium-241 is almost 10 times those of plutonium-239 and plutonium-240.

A common feature in Figs 2.1 and 2.3 is the dominance of the hazard from americium-241 in the first few thousand years. Later, plutonium-239 and 240 assume dominance in wastes from Once-Through and Equilibrium Cycles, whereas americium-243 is predominant in Thermal Recycle. The Fast Reactor Equilibrium Cycle gives the lowest total hazards at all times, usually by more than tenfold: this is to be expected, since the ratio of fission to capture is higher for any nuclide at the higher neutron speeds. Turning to Fig 2.4, the Once-Through Cycle creates the highest hazard at all times: Thermal

Recycle achieves some reduction through fissioning plutonium instead of disposing it. However, in this system, there is sufficient americium-243 first to dominate after americium-241 has mostly decayed away, and later, to give appreciable 'growth' of plutonium-239 (see the decay chain of americium-243 through neptunium-239 to plutonium-239 in Fig. 1.2). Since plutonium-239 has more than twice the ingestion toxicity (per Bq) of americium-243, the total hazard increases with this growth to give a secondary peak at about 10,000 years.

2.8 SUMMARY

Of the five systems discussed, the Once-Through Cycle is the only one which avoids the reprocessing necessary to recycle plutonium. As a consequence, however, the efficiency of uranium usage is very low and the plutonium arisings for disposal are over 20 times those of recycle routes. In fact, the present programmes for UK and French nuclear power, operated on a Once-Through basis, would each have accumulated hundreds of tonnes of plutonium early in the next century.

Recycling of plutonium drastically reduces plutonium in wastes, though markedly increasing americium arisings in the Thermal Recycle system, where uranium usage is still of low efficiency. The Fast Reactor Equilibrium Cycle produces the least long-lived wastes, mainly because Fast Reactors are more efficient in fission relative to neutron capture than Thermal Reactors: uranium usage is highly efficient. There are a range of combinations of Thermal and Fast Reactors which might be used to 'burn up' plutonium. With respect to the efficiency of uranium usage, this increases as such systems approach the Equilibrium Cycle of Fast Reactors only.

Using the potential ingestion toxicity as an index of waste hazard, the dominating nuclide from a few hundred years to a few thousand years after reactor discharge is americium-241 in all systems. A useful target in disposal concepts discussed later is therefore to design so that this nuclide has decayed away before it can make its way back to man.

It is perhaps worth re-emphasising that the above approach on toxicity is to derive a simple first indication of the relative hazard importance of long-lived waste nuclides or to compare possible cycles. If wastes were disposed in a guaranteed stable saline groundwater location, such as below the sea-bed, the relative hazards of the nuclides would be changed. A complex evaluation of possible ways in which the nuclides might reach man, e.g. by eating fish, would then be more relevant.

Appendix 2.1

EFFECTS OF STORAGE ON PLUTONIUM QUALITY

Factors affecting heavy nuclide arisings in all but the Once-Through Cycle are storage of plutonium both in spent fuel before the latter is reprocessed and also new fuel before charging it to the reactor. In storage, plutonium-241 (half-life 15 years) continues to decay to americium-241. If, for example, there is a delay of 15 years before reprocessing, half of the plutonium-241 is lost: on reprocessing, the americium formed follows the waste stream, the remaining plutonium-241 being recovered with the rest of the plutonium. The latter is then lower in quality because of the decay during storage. (Plutonium-241 is more fissile in reactors than plutonium-239.) On the other hand, if MOX fuel is stored for a long period before being charged to a reactor, americium-241 will replace some of the plutonium-241, again reducing the fuel reactivity, since its fissile cross-section is much lower than that of the latter: a higher level of americium-241 and heavier isotopes in waste streams will also result. Figures given in Table 2.1 assume PWR plutonium cooled for about 5 years (to decay out short-lived fission products) before reprocessing and stored for about 4 years before returning to a reactor.

Appendix 2.2

SOME VARIANTS IN PLUTONIUM REPROCESSING

A.2.2.1 <u>Dilution Reprocessing</u>

A variant on completely separate reprocessing of Thermal and Fast Reactor fuels, particularly when there are few Fast Reactors in existence, is to build separate Head End plants to break down and dissolve each type of fuel in nitric acid, then blend the two solutions together before a common Back End plant, which separates uranium and plutonium from wastes. This has been termed 'dilution' reprocessing. With only a small proportion of Fast Reactors, waste arisings are little different from those from the Thermal Reactors alone.

A.2.2.2 Breeder Recycle to Thermal Reactors

Yet another variant which could assist in keeping up the quality of plutonium recycled to Thermal Reactors is to blend breeder fuel in with Thermal Reactor fuel reprocessing, the core fuel from the Fast Reactors being processed separately.

A.2.2.3 The Reliability of Plutonium Recycling

When Fast Reactors make an essential contribution to the National Grid, the supply of plutonium must be very reliable indeed, otherwise they may have to shut down. A Fast Reactor reprocessing plant will probably therefore consist of multiple lines of small throughput: the elimination of any possible faults common to the lines will be vital. Many years of experience in operating prototype lines in exactly the manner required for commercial application is therefore desirable: this is a powerful reason for maintaining continuing development and operation of Fast Reactors and their associated fuel cycles.

Table 2.1

Plutonium Inputs, Uranium Usage Efficiencies, Arisings and Potential Ingestion Hazards of Long-Lived Nuclides in Wastes (per GW(E)yr).

System	Once-Through	Thermal Recycle	Fast Reactor Equilibrium
Plutonium Input (kgs PuE)	-	600	1,900
Efficiency of Uranium Usage (%)	0.5	1.0	50-80
Arisings of Nuclides as Wastes (kgs) (a)			
Pu239	180 (100)	6 (3)	7 (4)
Pu240	74 (200)	5 (10)	3 (6)
Pu241	44 (700)	2 (30)	<1 (<20)
Pu242	16 (0.5)	2 (0.006)	<1 (0.003)
Am 241	4 (60)	82 (1,000)	15 (200)
Am 243	3 (3)	78 (70)	<1 (1)

Notes:

(a) The hazard unit of the bracketed toxicity is one billion ALIs.

(b) Recycle plutonium is assumed to be of PWR quality and to have been stored for 4 years before loading.

(c) Plutonium inputs have been adjusted to PuE, their equivalent in reactivity to Pu239 in a Fast Reactor.

(d) In each cycle except Once-Through, reprocessing is assumed to lose 0.5% plutonium to wastes.

(e) Plutonium-238 is not listed, since it arises in only small quantities and its half-life is relatively short (88 years): however, its radiotoxicity can exceed that of plutonium-239 in 'young' spent fuel from Thermal Reactors (Elayi, 1987).

(f) Arisings are quoted at a few years after discharge from a reactor. Note that after 300 years, the earliest time plotted in the curves of Figs. 2.1 to 2.4, all of the plutonium-241 will have decayed to americium-241.

Figure 2.1: POTENTIAL INGESTION HAZARD (PIH) OF LONG-LIVED NUCLIDES FROM A 'ONCE-THROUGH' CYCLE

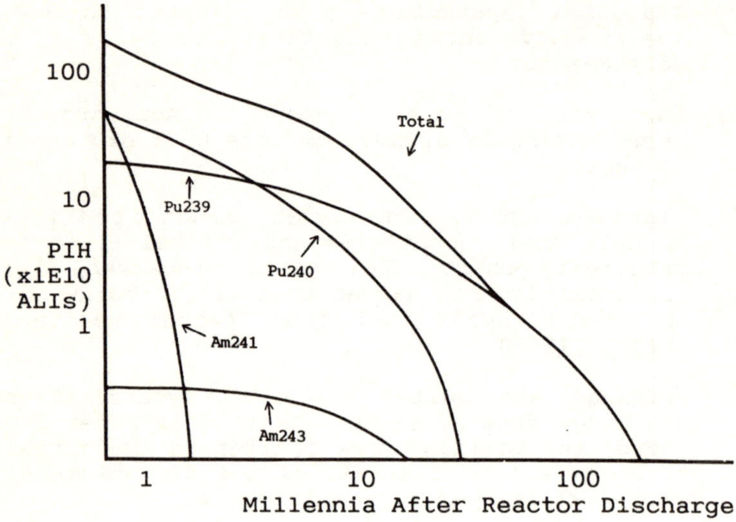

Figure 2.2: POTENTIAL INGESTION HAZARD (PIH) OF LONG-LIVED NUCLIDES FROM 'THERMAL RECYCLE'

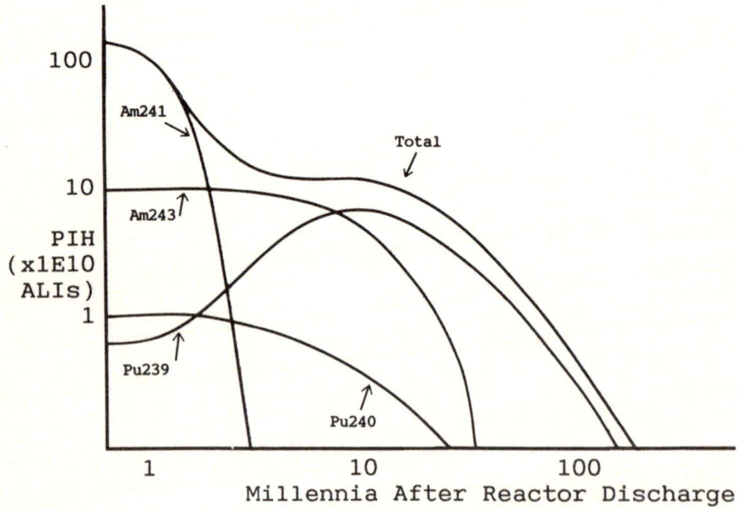

Figure 2.3: POTENTIAL INGESTION HAZARD (PIH) OF LONG-LIVED NUCLIDES FROM A 'FAST REACTOR EQUILIBRIUM' CYCLE

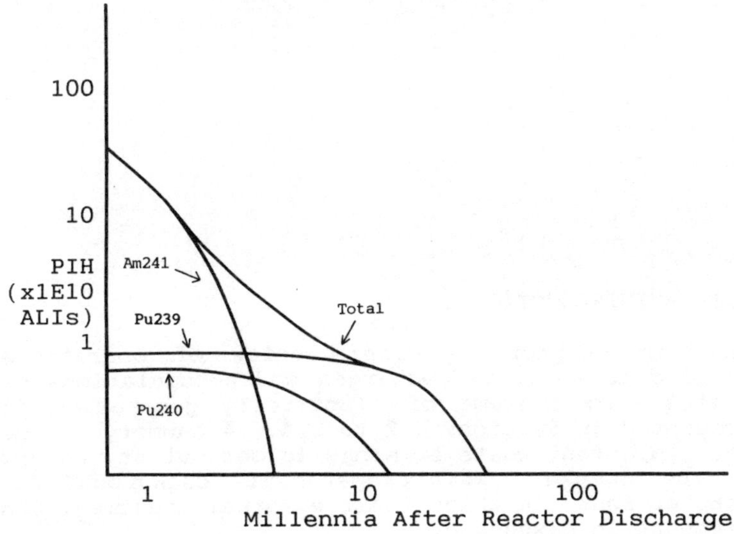

Figure 2.4: COMPARISON OF TOTAL INGESTION HAZARD (PIH) FROM VARIOUS NUCLEAR CYCLES

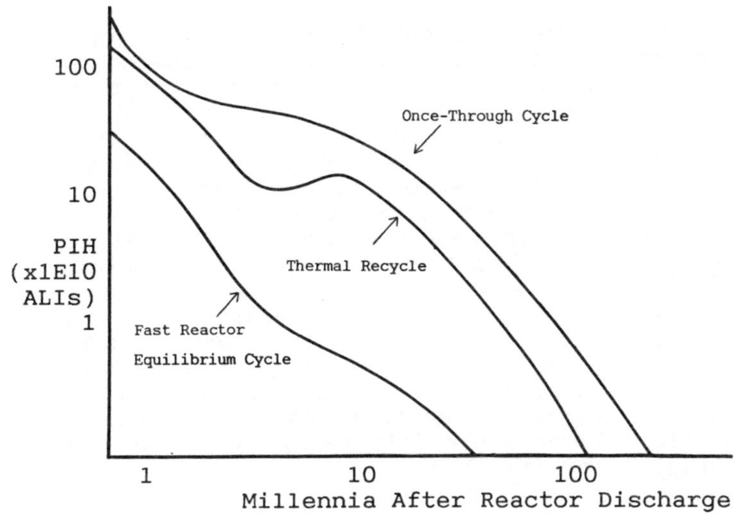

Chapter Three

ARISINGS OF POWER SYSTEM WASTES

3.1 INTRODUCTION

In this chapter, a brief review of operations leading to routine discharges and accumulations of wastes from systems of electricity generation is presented in Sections 3.2 to 3.4. A summary of the more important waste arisings is set out at the end of the chapter. This prepares the background for Chapter Four, in which their eventual treatment and disposal is discussed.

3.2 NUCLEAR POWER

3.2.1 General

In Chapter One, various types of power reactors were introduced, with the need to prepare several types of fuels - natural uranium metal for Magnox reactors, the dioxide of uranium enriched in the 235 isotope for Advanced Thermal Reactors and, for Fast Reactors, the mixed oxides of uranium and plutonium. The provision of the latter depends on the chemical reprocessing of spent reactor fuels. From the associated reactor and fuel cycle operations, basically two types of wastes are created. The first involves low concentrations of activity, but is of such high volume that storage is impracticable. These are then discharged immediately, as liquids through pipelines to sea or as gases to stack. The second type of wastes contains higher concentrations of radioactivity, but the volumes are sufficiently small that interim storage is practicable.

This chapter sets out to describe how the various forms of radioactive wastes arise, but before proceeding to an outline of the relevant operations, it is convenient to define the wastes which are stored awaiting treatment and disposal, in the groups commonly used in the literature (Flowers, 1984).

(a) Low Level Wastes (LLW) are solids that have radiation concentrations less than 4,000 Bq/gm alphas and 11,000 Bq/gm betas plus gammas, as packaged. These limits permit easy access; operators can lay hands on packages as necessary.

(b) Intermediate Level Wastes (ILW) are solids with higher activities than the above limits. During operations, they normally require to be shielded or to be handled with remote devices. The heat output and long-lived activity content, however, are usually relatively small, so that these properties are not normally important in disposal schemes.

(c) Plutonium-Contaminated Wastes (PCM) are solids mainly associated with Fast Reactor fuel fabrication, where the wastes have insufficient plutonium to merit attempts at further plutonium recovery. Though of low heat output and requiring little or no shielding, the long half-life of the activity necessitates careful design of disposal facilities to hinder the return of plutonium to the biosphere. PCM can be considered in many problems as a subset of ILW.

(d) High Level Waste (HLW) is associated with the bulk of the activity remaining after the recovery of uranium and plutonium during reprocessing. As initially produced, it is held in tanks in a nitric acid solution, known as High Active Liquor (HAL): the current intention is that HAL should eventually be vitrified. The resulting glass blocks emit significant heat and thus are known as Heat-Emitting Wastes, or HEW; this factor must be taken into account in disposal by underground burial. Further, their long-lived activity, which can vary according to the recycle systems described in Chapter Two, is sufficiently high that the containment of this can become an important consideration in the design for their disposal.

(e) Spent fuel, if disposed as such, presents both the problems of the above HEW glass and those of containment of appreciable concentrations of plutonium and americium, described in Chapter Two.

The above broad groupings refer to current properties of waste. If there is a long period of storage, the ensuing decay can alter the characteristics so that a given waste is downgraded to a lower group at the time of disposal, i.e. either ILW becomes LLW or HEW becomes ILW. The more important of such changes will be noted as individual wastes are discussed.

3.2.2 **Preparation of Fuel**

The raw fuel source for UK nuclear power is received from abroad as uranium ore concentrate at Springfields near Preston, where it is further refined to remove the radioactive decay products of uranium (see the decay scheme for uranium-238 in Fig. 1.1). The most important of these impurities are the beta-emitting nuclides thorium-234 and protoactinium-234, which are rejected via pipeline to the tidal stretch of the River Ribble. The alpha activity discharged from the site is about 100 times lower than the beta activity and arises mainly from uranium.

The purified nitrate of uranium is then converted to the tetrafluoride. Uranium for Magnox reactor fuel requires no enrichment; the metal bars, made via magnesium reduction of the tetrafluoride, are canned in Magnox alloy and dispatched to the appropriate power station. For AGR and PWR fuels, the uranium tetrafluoride is converted at Springfields to the volatile hexafluoride, then enriched in the 235 isotope at Capenhurst in Cheshire, as described in Appendix 1.4, by either the diffusion or centrifuge processes. Some of the uranium arriving at Capenhurst has been recovered from spent fuel at Sellafield: as a consequence, small quantities of the fission product technetium (which, like uranium, has a volatile fluoride) occur in process streams. However, this and uranium, which contribute the large majority of the activity in wastes, only give rise to trivial activity for disposal locally to the Rivacre Brook and the Meols sewage pipeline. The enriched uranium hexafluoride is converted to oxide at Springfields, followed by cladding in either stainless steel tubes (AGR and PWR) or zirconium (PWR). Overall, operations at Springfields yield only slightly radioactive solid wastes. Some of this has such low activity that it can be dumped at local council sites: slightly higher level active material is sent to the Drigg site in Cumbria (to be described later in Section 4.1.7).

Currently, plutonium fuels for the core of the Prototype Fast Reactor (PFR) at Dounreay are fabricated at Sellafield. These mixed oxides (MOX) of uranium and plutonium pass through many stages in the production line before their canning in stainless steel; inevitably there is some small fraction rejected as scrap. Part of this can be recycled; the residues constitute PCM waste. In future, the use of plutonium in commercial-scale reactors, either Fast or Thermal, will create larger arisings of PCM.

3.2.3 Power Operations

Wastes arise during power production in gaseous, liquid and solid form; the nature of these from the Thermal Reactors, Magnox, AGR and PWR, is broadly similar, as are the techniques for their management (Passant, 1985). Gases are routed by ventilation systems and then discharged to stack via high efficiency filters. Argon-41, produced in the shield coolants of Magnox and AGR reactors, passes through the filters and is the main component of activity released to atmosphere. However, levels discharged are commonly only about 1% of regulatory limits: corresponding estimates for the Sizewell B PWR are more than a factor of 10 lower than this. Nevertheless, in spite of these low-hazard discharges, there has been controversy over whether there is a higher incidence of leukaemia near nuclear power stations (as discussed further in Section 6.2.3).

Discharged fuel is normally stored in water ponds and some slight activity transfers to the water. These active liquid wastes are filtered, sometimes through ion exchange resins (operating like conventional water softeners), leaving some activity from tritium, sulphur-35 and other nuclides (fission products from Magnox stations, induced activation products from AGRs) in liquid discharges from the site. However, again levels are only a fraction of the regulatory limits, varying from 1% to 20%, according to site.

LLW arises from discarded operating materials such as tissues, plastic coverings, etc. Much of the ILW in Magnox reactors, AGRs and PWRs, is produced from sludges and the resins used to clean up reactor circuits and the fuel storage ponds. Components discarded after service within the reactor can have high induced cobalt-60 activity; these are stored in the reactor vaults to become much lower activity ILW by the time the vaults are emptied when the reactor is decommissioned.

Current designs of Fast Reactors depend on cooling by sodium (see Section 1.2.3): ILW operational wastes can therefore be rather different from those of Thermal Reactors. Only a small amount of activated component debris is generated, plus an occasional spent trap for cleaning up circuit sodium. Items with adhering contaminated sodium, including spent fuel, are steam cleaned, giving rise to a slightly contaminated aqueous effluent. They are then stored under water in ponds as for the Thermal Reactors above, giving rise to similar wastes on cleanup of the pond water.

3.2.4 Fuel Reprocessing

The operations involved in reprocessing spent fuels are often grouped for convenience in description into Head End and Back End operations (Section A.2.2.1). The first of these includes operations from storage of spent fuel as received from reactor sites to the breakdown and dissolution of the fuels in nitric acid. Back End operations comprise the remaining operations required to separate out the uranium and plutonium and treat resulting wastes.

Temporary storage of spent fuel at Sellafield, awaiting reprocessing, is under water in ponds like those at the reactor sites above: similar cleanup resin wastes occur. Corrosion of Magnox fuel cladding is minimised by control of the chemistry of the pond water, but immersion for long periods releases significant activity, in particular caesium-137. In the 1970s, delays in reprocessing with corresponding longer pond storage caused increases in activity in pond purges to the pipeline to sea. Special cleanup units to pick up caesium in the ponds have now been installed; with these and the resumption of normal reprocessing, discharges to sea have decreased dramatically. By comparison, cladding of stainless steel or zirconium corrodes much more slowly than that of Magnox fuel, so that ILW cleanup wastes from AGR, PWR and Fast Reactor fuel ponds are of lower volume than those from Magnox fuel ponds.

In reprocessing Thermal and Fast Reactor fuels, the initial operation is to break down the grid which holds the fuel rods together. ILW wastes from this consist of small sections of metal of high induced activity as in the reactor component debris above, e.g. from cobalt-60 in stainless steel (see reaction (E) in Section 1.2.3), but with only a small component of long-lived activity. There are differences in the next step between Magnox and other fuels.

Magnox cladding is 'peeled' away from the fuel by pushing through cutters; about 1% of the fuel adheres to the Magnox fragments. Up to the present, these fragments have been stored as wastes under water, becoming partly converted to oxide sludges through corrosion. Recovery from the water and encapsulation in cement, as detailed in Appendix 3.1, is then scheduled. For future operation, the cladding will be encapsulated in cement as it is peeled off the Magnox fuel. Effectively, a similar waste product eventually accrues as for sludges, i.e. drums containing cladding wastes with a little adhering spent fuel, encapsulated in cement. These Cemented Magnox Swarf Wastes (for convenience designated here as CMS) are commonly classified for handling purposes as ILW: for disposal purposes, their appreciable plutonium content suggests their consideration as PCM.

AGR and PWR fuel pins are chopped up into short lengths into a basket with no separation of fuel from canning material; the fuel is then leached into solution by nitric acid, leaving the cladding residues undissolved in the basket. These ILW residues, unlike the Magnox cladding wastes, have no significant amount of fuel attached to them; they are, however, similarly encapsulated with cement in drums. Fast Reactor fuel will be similarly chopped up and leached, leaving stainless steel can fragments in the dissolver basket.

After decladding, there is a considerable similarity in the reprocessing of all fuels, though Magnox and Oxide fuels will continue to be treated at Sellafield in separate plants, B205 and THORP respectively. Boiling with nitric acid continues until the fuel is completely in solution; some radioactivity enters the gas phase at this stage. While some gases are retained by the use of scrubbers, others such as krypton-85 are discharged to the stack; activity levels are well within regulatory limits (BNFL, 1986). The cooled solution then begins the Back End stage by being contacted with a solvent (tributyl phosphate, diluted with odourless kerosene) to extract plutonium and uranium, leaving a high-active 'raffinate' aqueous waste solution from this first cycle of extraction. After concentration, this forms the preponderance of HAL, which is stored to await vitrification (described further in Appendix 3.1). Varying the acidity of the aqueous solutions allows extraction of the uranium and plutonium back from the solvent; re-extraction into solvent leaves behind more liquid wastes. These secondary cycle raffinates of intermediate activity levels are stored in tanks for several years to decay away much of the short-lived nuclides, such as ruthenium-106, before discharge by pipeline to the sea. Recently, extra treatment plant has been installed to reduce long-lived activity still further before discharge. For an excellent account of measures to reduce the discharge of radioactivity from the Sellafield pipeline, see Howden (1988). In a later extraction cycle, at carefully controlled concentrations and acidity, separation of uranium from plutonium can be brought about, if desired. However, in future flow-sheets for Fast Reactor fuels, such as the 'Nelson Circuit', such separation may be avoided, thus reducing the potential for theft of bomb-making materials (Jones, 1987).

3.2.5 Decommissioning

Nuclear fuel preparation plants produce little radioactive waste not discharged as part of normal production operations. The only radioactive operation remaining after operation ceases is therefore the decontamination of facilities. This produces fairly low-active liquids of little significant volume or total activity relative to that from the operational period. The structures after decontamination would be only very low activity LLW. Reprocessing plants tend to become more contaminated than fuel preparation plants, because of the higher complexity of the chemical operations involved. Nevertheless, the feasibility of decommissioning such a plant has already been demonstrated at Dounreay, where equipment in building D1206 was stripped down after the reprocessing of spent fuel containing highly enriched uranium ceased. Refurbishing was then carried out, so that plutonium spent fuels are now being reprocessed there. This example can therefore be used to deduce the arisings of waste from the removal of equipment and structures and the level of liquid waste from decontamination in future larger-scale plants.

It is currently proposed that nuclear reactors would be decommissioned in three stages.

Stage 1 would consist of removing the fuel from the reactors and then disconnecting the control systems, so as to prevent any further operation of the system.

Stage 2 would then involve the removal of peripheral components which are substantially unaffected by radiation, e.g. turbines, secondary circuits, laboratories, etc.

Stage 3 would then tackle the reactor structure itself, which would be significantly affected by neutrons during power operation: since much of the activity induced in these structures is relatively short-lived, e.g. cobalt-60, the dose to operators cutting up the structure is greatly reduced by waiting several decades or even a century before carrying out this stage. A judgement therefore has to be taken between radiation dosage to operators and the environmental impact of leaving shutdown reactors dotted throughout the UK. On the other hand, an attractive option, discussed further later, could be to site reactors where they could be left indefinitely after Stage 2.

Wastes that arise from Stage 1 of decommissioning can be considered as a slight continuation of on-power reprocessing; separate itemisation of such wastes would not really be necessary. At Stage 2, wastes are mostly very low-active LLW, but are generally of a large volume, e.g. slightly active concrete shielding. Stage 3 wastes are of smaller volume, mainly ILW, but with a high activity unless demolition is delayed for decades. Currently, an exercise in decommissioning a reactor is being undertaken on the small prototype AGR at Sellafield. In the US and West Germany, commercial-scale reactors will soon be decommissioned at Shippingport and Karlsruhe respectively. The AGR and the overseas reactor work, together with numerous desk studies, will provide data to provide estimates of types and volumes of wastes from the commercial stations decommissioned in the future.

A special extra problem arises with Fast Reactors, where there is about 1,000 tonnes of sodium coolant in the circuits of a commercial reactor, mainly contaminated with traces of caesium-137 which has not been removed by sodium cleanup traps. In principle, this might be reused in later reactors, but transport problems might be serious if the new reactor were not close at hand. A first step towards disposal might be to burn the sodium to oxide, then dissolve this in water. Neutralisation with acid might then enable treatment with special resins to pick up the main activity from caesium, as in Sellafield pond treatment.

3.2.6 Long-term Storage of Wastes

Two arguments for the long-term storage of nuclear waste are (a) more time for the development of improved methods of disposal, and (b) the simplification of disposal caused by substantial decay of radioactivity during storage. Such potential benefits must be balanced against the additional radioactive operations of storage and the associated extra costs. Further, a holdup period of several decades raises questions as to the ability of the original equipment to handle waste packages after such a long time. Expensive materials, such as stainless steel to provide resistance against corrosion, may therefore be necessary for both equipment and storage containers.

Unlike the storage under water of spent fuel at most Magnox stations, the spent fuel at Wylfa is held in dry storage, the heat being removed by air.

Corrosion is slower than in pond storage, so that the fuel can be held for longer periods without reprocessing (Jones, 1987). Dry storage is also a possibility in the future for AGR and PWR spent fuels, where several decades may be an acceptable holdup before reprocessing or disposal: both stainless steel and zircaloy cans resist corrosion well under such conditions. Secondary wastes from this form of storage are insignificant.

As will be seen later, two major features affecting safety after disposal are the toxicity of the waste in possible ingestion after migration into groundwaters and the heat output from packages. Long-term storage before disposal is unlikely to assist in reducing toxic hazards since the storage time is short relative to the duration during which nuclide migration takes place after disposal, i.e. shorter-lived nuclides will decay away before migrating to the biosphere, whereas the longer-lived nuclides will be virtually unaffected by a storage period of a few decades. On the other hand, there is a considerable fall in heat output from HEW during the 50-100 years after reactor discharge (see Appendix 4.1 and Fig. 4.1) and a corresponding reduction in disposal problems, e.g. a reduction of excavation in tunnel emplacement of vitrified waste and a corresponding radiation dose reduction during disposal operations (Burton, 1981). To a lesser extent, in ease of handling, some benefit may also be gained from the storage of cemented cladding wastes, but there would seem to be no significant technical advantage in holding up the disposal of other cemented wastes.

3.3 POWER FROM COAL

3.3.1 Preparation of Fuel

The volume of water pumped out in draining mines varies considerably, from 0.1 to 8.1 (average 2.3) tonnes per tonne of saleable coal (Comm. En. Env., 1981). This contains chemicals leached from the ground through which the water has percolated. Elements such as iron may be oxidised to form brown sludges when these waters are discharged on the surface. In the separation of spoil from coal, modern mechanical systems make it cheaper to do the washing above ground rather than down the mine;

typically, about one tonne of spoil for every two tonnes of coal then arises. Some recycle of the wash water is possible, but there is still a large discharge of waste water containing dissolved solids and suspended solids. Some 5 to 30% of the sulphur present in coal as iron sulphides (pyrites) can be washed out. Spoil is usually heaped near the colliery and further leaching by rainwater transfers chemicals to local water courses or groundwater. At 4 p.p.m. uranium in this spoil (the average concentration in the Earth's crust) about 7 tonnes of uranium per GW(E)yr of electricity production are dumped in spoil heaps. Most of the radioactivity in this uranium comes from uranium-238 and its 8 alpha and 6 beta-gamma descendants (see Fig. 1.1). The above 7 tonnes of uranium are then equivalent to about 8 TBqs (200 Ci) of alphas and 6 TBqs (160 Ci) of beta-gammas.

3.3.2 Operational Wastes

The bare outlines of the combustion of coal to produce power have already been given in Section 1.3. The stack gases there described contain enormous quantities of carbon dioxide, nitrogen oxides and sulphur dioxide. At a lower output within these gases are discharged fly ash particulates which are too fine to be picked up by the filters. The composition of these has some similarity to the mineral residues originally in the coal: consequently, the following elements are often found in a highly toxic form: selenium, vanadium, lead, mercury, cadmium, arsenic and beryllium, some of which are carcinogenic, as discussed later in Chapter Five.

The hydrocarbons in the stack gases are numerous, varying with the coal source: perhaps the most important one is benz-alpha-pyrene, which is a known carcinogen (Comm. En. Env., 1981). Radioactivity from stack discharges, mainly of radium and thorium, depends on the mineral content of the coal. At 1.4 p.p.m. uranium in coal (Ewart, 1983) about 5 tonnes per GW(E)yr would be discharged, partly to stack. Using similar calculation to Section 3.3.1, this corresponds to about 6 TBqs (150 Ci) of alphas and 4 TBqs (100 Ci) of beta-gammas discharged per GW(E)yr. Future coal-fired power stations will come under an EEC directive of 1984 requiring best technology to be used against air pollution without excessive cost increases (Barrett, 1986).

Large volumes of liquid waste arise from power stations because of the common practice of 'lagooning' ash from the furnaces and gas scrubbers. This entails transferring ash to the preferred disposal point as an aqueous slurry. After settling of the solids, water is allowed to discharge, usually to the local surface or groundwater flow. The soluble part of the ash, containing boron, radionuclides and toxic heavy metals, therefore migrates with the leach water.

Part of the ash produced during power production is sold for such purposes as landfill and the manufacture of building materials. The same toxic elements as in the stack gas 'fines' as mentioned above are therefore widely distributed throughout the UK. The unsold ash is still too voluminous to justify the cost of transporting it more than a few miles. Consequently, large areas near coal-fired power stations are occupied by the lagoons described above. The leaching of toxic trace elements from these ash dumps is admitted to be a main environmental concern (British Coal, 1987).

3.3.3 Decommissioning

The demolition of coal-fired power stations is not appreciably different from that of many industrial buildings in general. This is the case, too, with the surface plant at collieries. The respective volumes of waste are small relative to those arising during the operational life of the system.

There are three main problems after shutdown of plant operations. The first of these arises from the vast quantity of dumped spoil and ash which continue to release chemicals by leaching. Usually, an attempt is made to ameliorate the physical aspect of the problem by landscaping the dumps or grassing them over. The second problem is subsidence caused after coal and spoil have been excavated. Modern methods of shearing at the coal-face produce controlled subsidence, the strata collapsing behind the shearer as it proceeds forward: the areas of land affected are considerable. Thirdly, after abandoning a mine, it gradually becomes flooded; eventually, there is an outflow of mine waters containing iron and other metals in solution. On reaching the surface, these can be oxidised and precipitated as hydroxides. The pollution can be sufficiently serious that fish are exterminated in nearby streams and rivers, as discussed further in Section 6.3.4.

3.4 FUSION POWER AND RENEWABLE SYSTEMS

3.4.1 Preparation of Fuels

The fusion of deuterium and tritium to produce power was outlined in Section 1.4.2. Equation (F) gave the basic fusion reaction, while Equation (G) showed how neutrons can create the tritium part of the fuel from lithium-6. Usage of this isotope leaves a residue of lithium depleted in lithium-6; however, this residue is not radioactive and so could be sold for conventional lithium applications. The separation of the tritium from lithium and then from the by-product helium, yields the latter as a waste product. However, its contamination therein by tritium would be slight and venting to atmosphere acceptable. For the deuterium part of the fuel, a heavy water (deuterium oxide) plant of the scale currently operating in Canada could provide adequate supplies for many fusion reactors; the waste arisings of such a plant, mainly from the slight discharges of hydrogen sulphide to the atmosphere, would not be significant. The more advanced fusion reaction using deuterium only (Equation (H), Section 1.4.2) would be even easier to supply with fuel.

Renewable processes do not, of course, use fuels. However, indirectly, appreciable quantities of fuel may be required in the preparation of constructional materials, e.g. the cement involved in the construction of a Severn Barrage may well run into millions of tonnes - this means that considerable quantities of heat must be supplied to decompose limestone to calcium oxide in the manufacture of the cement: the secondary wastes discharged in so doing are necessarily considerable.

3.4.2 Operation

Only fusion power consumes fuel. The reactions are in the gas phase, however, and only tritium is radioactive. Its discharge to atmosphere should be relatively innocuous (Section 1.4.2). Periodic discard of activated structural materials are appreciable (Davis, 1987) being perhaps double the volume from a PWR of equivalent power (Tables 3.1 and 3.3).

3.4.3 Reprocessing

The only reprocessing occurs with the fusion systems. A helium purge must be maintained to control its accumulation in the reaction zone (see Equations (F) and (H) of Section 1.4.2). Proven separation methods are known - one procedure could be to absorb deuterium and tritium on uranium as a hydride. After venting off the helium, the deuterium and tritium (where present) could be regenerated by heating and the uranium used in a further cycle. Alternatively, the trapping of hydrogen isotopes as water, after combination with air in contact over a catalyst, is a straightforward process (in fact, it is a routine procedure in deuterium recovery in the Winfrith SGHWR). After venting the helium to stack, the water could be decomposed to recycle the hydrogen isotopes. Any release of tritium, say in trace concentrations with vented helium, should be relatively innocuous (Section 1.4.2).

3.4.4 Decommissioning

Renewable systems require only conventional demolition at the end of their operating life. In fusion power, however, the deuterium-tritium reaction gives rise to neutrons which can activate structural materials. Since the total number of neutrons entering these materials will be less than in fission reactors of the same power output, there will be correspondingly less induced activity during disposal operations (Hancox, 1987).

3.5 SUMMARY AND ESTIMATES OF WASTE ARISINGS

3.5.1 General

The review above of the ways in which wastes arise from the various power systems is summarised below. Such features are noted, not only because of their scale or hazard, but also where they have been the subject of controversy. In the quantification of the arisings for 'one-off' operations, such as decommissioning, a 30-year system life has been assumed in most cases; the corresponding arisings are then averaged over this period. Most of the data used in the compilations of Tables 3.1, 3.2 and 3.3 come from Beckmann (1976), Cohen (1983), Comm.En.Env. (1981), Day (1985), Greenhalgh (1980) and HSC (1978).

3.5.2 Nuclear Power

In fuel preparation, the main wastes come from PCM in the recovery of plutonium and the fabrication of MOX fuels. During power operation, materials used by operators are discarded as LLW, and ILW is created from scrap equipment and by cleaning up reactor circuits. Fuel pond storage and its associated cleanup creates more ILW. Reprocessing produces a variety of LLW from operating materials and ILW from fuel breakdown and scrap equipment. HLW arises as a nitric solution containing the bulk of the fission products and unwanted heavy isotopes such as americium and neptunium: long-term storage of spent fuel creates little waste. The main waste streams are shown in Table 3.4.

Estimates of nuclear wastes for unit power generation of 1 GW(E)yr have been listed in Table 3.1. Both volume and mass are given, even though one form may seem inappropriate, e.g. the mass of HAL as liquor. This is to facilitate comparison with the non-nuclear arisings listed in the following Table 3.2. It must be pointed out, too, that net spent fuel arisings and reprocessing wastes are, in effect, alternatives - they should not be taken together in the total operation of a nuclear system, even with a long period of storage before reprocessing.

3.5.3 Coal-Fired Power Wastes

Large volumes of solid wastes occur here through coal cleaning and fly ash residues: these must be dumped above ground not far from their origin. Toxic components of liquid wastes from the leaching of dumps of the above can be estimated by assuming they are leached out in a few years, so that their rate of removal from an established dump is equal to the content in new wastes being added. (A lower leach rate merely means the hazard is spread over a longer time period.) Gaseous discharges of carbon dioxide are huge; emissions of the inorganic oxides of sulphur and nitrogen are also massive. Future processes to reduce undesirable gaseous emissions, described in Section 1.3, will inevitably produce large volumes of non-gaseous wastes (Barrett, 1986). The arisings of the polycyclic hydrocarbons and trace toxic elements including radioactive components in the stack gases are not negligible. About 5 tonnes each of arsenic and mercury, for example, are discharged to atmosphere from a 1 GW(E) coal-fired power station each year in the UK. It is uncertain what effect new equipment and processes such as FGD and FBC will have on trace element discharges (British Coal, 1987) but the secondary environmental effects will be severe (Longhurst, 1987). The alpha activity in stack gases is, in fact, similar to that in LLW from nuclear power. A summary of the waste streams is outlined in Table 3.5; quantitative estimates for some of the above wastes are listed in Table 3.2.

3.5.4 Fusion Power and Renewable System Wastes

Fusion power creates less complex mixtures of wastes than fission power, because of the absence of fission products and heavy nuclides and because of the simplicity of such reprocessing as is necessary. Perhaps a half of the fusion power wastes are encountered during decommissioning; a rough figure for this may be derived in relation to the corresponding fission reactor decommissioning. Davis (1987) suggests 'a few hundred tonnes' per GW(E)yr during operation; much of this is activated steel and so will probably be classed as ILW.

The renewable power systems produce secondary wastes during the preparation of structural materials. Inherently, there is little or no waste during operation. Some rough estimates of these waste arisings are listed in Table 3.3.

3.5.5 Comparison of the Physical Arisings of Power System Wastes

The stack discharges from nuclear power stations are of insignificant levels relative to those from coal power. LLW arise mainly during operation and are several hundred times less in volume than the spoil from coal mining, over fifty times less in quantity than the fly ash and constituents of stack gases from coal combustion. Broadly speaking, for each GW(E)yr of power, the radioactivity in LLW is ten times less in alphas and about the same in beta-gammas as that arising in spoil and operational discharges from coal-fired power (see Table 3.1 Note (c) and Table 3.2 Note (d)). ILW is very roughly a factor of 10 lower in volume than LLW. The arisings of HEW or spent fuel are lower still: they are of a similar level to the arisings of toxic trace elements in the spoil and fly ash dumps and in the stack discharges of coal power.

The ILW arisings from fusion are similar in volume to those from fission power; there are no HEW from fusion. As for the other systems - nuclear and coal - which also require power production via heat, fusion could in principle be used in CHP, when, effectively for the same energy production, less wastes would be produced. Wastes from renewable systems arise mainly during decommissioning. Wind and wave systems give small arisings of conventional wastes - even lower than LLW from fission power. The case of tidal power is interesting in that large arisings of waste could be attributable to decommissioning. However, it might reasonably be argued that the demolition of a tidal barrage need not be considered - the operation would be costly and there would probably be no better place to put the rubble than where it was built. This invites a similar question for nuclear reactors, particularly with respect to the value of dismantling them beyond Stage 3 of decommissioning. Such a possibility could affect the original choice of site, as elaborated further in Chapters Four, Nine and Ten.

Appendix 3.1

CONDITIONING OF NUCLEAR WASTES

A.3.1.1 <u>General</u>

Additional treatments may be carried out on the basic liquid and solid wastes from nuclear operations in order to provide forms more suitable for storage, transport and disposal, as outlined below.

The normal objective is to provide a solid form capable of reliably retaining radionuclides over long periods, usually at least a century. Chemical forms which break down under gamma rays can only be used with low concentrations of radioactivity. For example, the water molecules in cement are broken up by radiation: hydrogen and oxygen gases are released and the cement crumbles. To encapsulate a given amount of radioactivity at an acceptable dilution in cement therefore produces a relative large volume of cement blocks. On the other hand, glass and ceramics are highly resistant to breakdown by radiation; activity concentrations can be relatively high and resulting product volumes relatively small.

A.3.1.2 Solids

Although other materials, such as bitumen, have been used in various countries overseas, cement has been preferred in the UK for encapsulating LLW and ILW (Howarth, 1987). A mixture of waste and cement is poured to set in steel drums of capacity either 200 litres (about the size of a common oil drum) or 500 litres. The drums are fabricated from a stainless form of steel if a long period of storage is envisaged before disposal. Much of the ILW arisings described earlier will be encapsulated in this way; some LLW and all PCM may also be cemented for ease in handling or transport. The encapsulation of Magnox swarf in cement will, of course, cease when Magnox reactors are finally shut down. Perhaps 20,000 m^3 of CMS will have been produced by this time. HEW and spent fuel could be overpacked, e.g. surrounded with thick extra layers of metal, both to facilitate handling during disposal and to inhibit ingestion of long-lived activity thereafter (Burton, 1981; KBS, 1983).

A.3.1.3 Liquids

For the solidification of HAL at Sellafield, a borosilicate glass will be prepared and cast into stainless steel containers by the continuous AVM process developed in France. This form of glass has a low leach rate in typical groundwaters and thus can be considered as the first barrier to migration of the enclosed nuclides after burial. The vitrification process is quite complex, operating at over 1,000 °C and generating secondary forms of waste from nuclides such as ruthenium which are in a volatile form at these temperatures. Higher temperatures are being proposed in other countries for the preparation of products simulating natural minerals of even lower leach rates than borosilicate glass. Useful reviews of the many variations of the basic idea of waste encapsulation are given by leading proponents Ringwood (1978) and Roy (1982). However, the additional complexities and costs involved make it unlikely that the UK will change processes, particularly as the current process provides sufficient safety against leaching in probable designs for disposal.

The loading of waste nuclides in glass is not only determined by the chemistry of the resultant waste matrix, but also by the heat output from the waste. Consequently, if reprocessing is required quickly after discharge of spent fuel from a reactor, as in the Fast Reactor Equilibrium Cycle described in Section 2.5, then either more matrix must be produced with a lower waste concentration or expensive tanks must be provided to hold the HAL for up to 5 years before vitrification to reduce the heat output by decay. This latter variation has been proposed for processing European Fast Reactor fuel at Dounreay (UKAEA, 1985). On the other hand, some existing long-stored HAL at Dounreay is scheduled for cementation; this has been preferred to vitrification because of the relatively small volumes requiring treatment, which are felt not to justify the more sophisticated and expensive process of vitrification. Moreover, the original fuel was about 80% uranium-235, so that most fissions came from the latter and only a relatively small proportion of absorption by uranium-238 took place, yielding little long-lived plutonium and americium.

Table 3.1

Physical Aspects of Nuclear Power Waste Arisings Equivalent to 1 GW(E)yr Operation (a),(b)

Waste Class	Arising per GW(E)yr (m^3)	(tonnes)
LLW		
Magnox	2,500(400)	5,000(800)
AGR	2,000(200)	4,000(400)
PWR (c)	2,000(30)	4,000(60)
FBR	2,000(100)	4,000(200)
ILW		
Magnox	300(450)	600(900)
AGR	70(300)	140(600)
PWR	100(50)	200(100)
FBR	100(100)	200(200)
PCM	140(0)	300(0)
HAL	20	30
HEW	5	15
Spent Fuel		
Magnox	12	200
AGR	5	50
PWR	3	30
Stack Discharges	vs	vs

Notes: vs denotes very small.

(a) As conditioned for transport and disposal.

(b) Bracketed figures give decommissioning arisings averaged over a 30-year plant life.

(c) For example, from the definition of LLW (Section 3.2.1), PWR LLW wastes would contain <2 TBqs (50 Ci) alphas and <40 TBqs (110 Ci) beta-gammas per GW(E)yr.

Table 3.2

Physical Aspects of Coal-Fired Power Waste Arisings Equivalent to 1 GW(E)yr Operation

Waste Class	**Arising per GW(E)yr** (m^3)	**Arising per GW(E)yr** (tonnes)
Mine Spoil	1,000,000	2,000,000
Arsenic (a)	3	5
Beryllium	3	6
Chromium	100	200
Nickel	40	80
Uranium (d)	4	7
Fly Ash	300,000	300,000
Stack Discharges		
Particulates (b)		300,000
Carbon Dioxide		13,000,000
Carbon Monoxide		20,000
Nitrogen Oxides		30,000
Sulphur Oxides		100,000
Arsenic (c)		5
Lead		1
Mercury		5
Uranium (d)		5

Notes:

(a) Rough figures, based on the average composition of the Earth's crust, for carcinogens: there are, of course, very large arisings of more innocuous elements, e.g. calcium.

(b) Particle size less than 20 microns.

(c) Arsenic to uranium here are some typical trace toxic components of the combustion gases: others include polycyclic hydrocarbons, beryllium, cadmium, carbon disulphide and hydrogen sulphide.

(d) Corresponding radioactivity levels (see Sections 3.3.1 and 3.3.2) in combined spoil and stack discharges are about 14 TBqs (380 Ci) alphas and 10 TBqs (270 Ci) beta-gammas.

Table 3.3

Physical Aspects of Waste Arisings Equivalent to 1 GW(E)yr Operation of Fusion and Renewable Power Systems (a),(b)

Waste Class	**Arising per GW(E)yr** (m^3)	(**tonnes**)
Fusion	200	400
Tidal	100,000	200,000
Wave	1,000	2,000
Wind	1,000	2,000

Notes:

(a) Decommissioning wastes are averaged over a 30-year period, except for a 100-year period for tidal power.

(b) Wastes from decommissioning form the bulk of arisings in all cases except fusion: the value of decommissioning a tidal barrage is, of course, uncertain.

Table 3.4

Principal Waste Streams from a Nuclear Power System

Fuel Stage	Preparation	Power Operation	Storage	Reprocessing		
Primary Wastes	PCM (Plutonium Fuels only)	Component Debris (ILW)	Spent Fuel	LLW	Cladding Resins (ILW)	HAL
Treated Waste	Cemented PCM	Cemented Debris	Overpacked Fuel	Packaged LLW	Cement Matrix	Vitri-fication

Note: Spent Fuel Storage and Disposal is an alternative, not an additional stream to Reprocessing.

Table 3.5

Principal Waste Streams from a Coal-Fired Power System

Fuel Stage	Mine Drainage	Excavation		Combustion				
Primary Stream	Surface Discharge	Spoil		Fly Ash	Stack Discharge			
Disposal Form	Ditto	Washings	Spoil Dump	Bricks or Dump	Carbon Dioxide	Acid Gases (a)	Trace Inorganics (b)	BAP (c)

Notes: (a) Acid gases are oxides of nitrogen and sulphur (limestone traps can change some of the stream to be treated like fly ash).

(b) Trace inorganics are toxic metals in low concentration.

(c) BAP are polycyclic hydrocarbons from incomplete combustion.

Chapter Four

THE DISPOSAL OF LIQUID AND SOLID WASTES

4.1 NUCLEAR POWER

4.1.1 General

In the previous chapter, the basic arisings of wastes from routine operations of nuclear power systems were outlined. Of the two general types of waste described in Section 3.2.1, the large volumes of gases and low-activity liquids cannot be stored and so are discharged immediately. An approach commonly known in the disposal of general hazardous wastes as 'dilution and dispersion' can be used, i.e. unacceptable local concentrations on release are avoided, such as by mixing with other effluents or through efficient dispersal at a safe point, or by a combination of the two.

Dispersal of gaseous effluents from nuclear and other power systems is efficiently carried out through tall stacks. It is therefore not discussed further in this chapter, though the impact of such discharges to atmosphere is assessed in later chapters. Pipeline discharge of low-activity liquids, e.g. at Sellafield, has, however, been the subject of much controversy and the manner of discharge is followed up here, together with options for the disposal of the second type of nuclear waste mentioned in Section 3.2.1, the stored wastes. These options mainly fall into the category of 'concentration and containment', as practised in general hazardous waste disposal.

Not so many years ago, a design engineer investigating disposal schemes would only have to assess features such as cost, technical feasibility and the constraints of appropriate regulations. Having prepared a short list of the more promising concepts, he would home in on a preferred scheme and set out the reasons why it was superior to alternatives. Consultation with regulatory authorities would go hand in hand with the evolution of the preferred scheme.

However, in the last few decades, protest groups have been set up against many major projects. This has been particularly so on issues such as motorway siting, national airports and nuclear power, where a national requirement can be met from a relatively small area. The local inhabitants receive only a minute fraction of the overall benefit but suffer the majority of the inconvenience and nuisance. Protest groups have become progressively more organised and widely influential, e.g. Greenpeace and Friends of the Earth (FOE) make representations on environmental issues at both national and international level. Foreign governments, seeking political leverage, have also 'joined the band-wagon'. Such opposition presents both technical and non-technical 'evidence' against nuclear waste disposal concepts. In this chapter, the more important of these will be mentioned, so that in the eventual comparison of overall systems we can decide whether such matters are relevant.

4.1.2 Technical Objectives

Many tonnes of paper have passed between the desks of bureaucrats in trying to decide the policies, strategies and criteria to be adopted in the management of nuclear waste. Instead of a careful comparison of properly drawn-up concepts, regulations have been set out after long-winded departmental procedures which inevitably involve 'mental designs' and legislative semantics. Here, we will return to the design comparison approach, first listing a set of guidance points derived from official publications, mainly relevant to the disposal of solid wastes.

The first guidance point concerns the possibility that, after disposal, waste may be inadvertently disturbed; a small number of persons may then be irradiated from outside the body or inhale radioactivity. A so-called 'intrusion barrier' of at least one metre of concrete or its equivalent has therefore been recommended for land burial of most types of waste except LLW (DOE, 1986). A second point arises because, in many concepts, leaching of radioactivity from packages may eventually occur due to circulating water under the land or in the sea: transfer of this activity is then possible to the biosphere. The UK DOE stated in 1984 that 'the site should be chosen and the facility designed so that the risk or probability of fatal cancer, to any member of the public, from any movement of radioactivity from the facility, is not greater than one in a million in any one year.' An approach arising from this is that a 'multi-barrier' system should be constructed (Flowers, 1984). Examples of successive barriers to migration of radioactivity are low-leach forms of waste, a thick container of corrosion-resistant material and a disposal location of low permeability to surrounding water. It is worth noting here that the 'multi-barrier' criterion is not a normal requirement in general waste disposal, where one satisfactory barrier is considered adequate. The multi-barrier requirement is overtly more to alleviate public anxieties over nuclear waste. Cynically, it might be said to be an attempt by nuclear organisations to increase profits through more complicated operations. A third point is that low-probability events such as earthquakes, volcanoes, etc. should be considered in the design specification; however, it has been found in studies to date that such events are unlikely to affect disposal arrangements in or around the UK, e.g. Burton (1981).

To the above 'official' guidelines we shall add one other. There is an apocryphal story of a designer reviewing the work of a new recruit. 'You've forgotten the KISS.' 'What's that?' said the recruit. 'Keep It Simple, Stupid' was the reply. If there is any technology where there is a general trend towards greater complexity, it is nuclear power. This is in spite of the corresponding increase in the chance of accidents and the difficult and costly cleanup of radioactive contamination that usually follows. Here, we will ignore concepts of waste disposal by rockets into deep space and keep our feet firmly on the ground!

Following the general trend of world research, there are at present three broad types of location for the practical disposal of radioactive wastes - to the sea, under the sea-bed and under land. Disposal under land is considered here to include disposal under the sea-bed with access from land. This is reasonable, since the general engineering problems are similar to those involved in disposal beneath land and markedly different from those involved in disposal from ships or jackup platforms out at sea. We shall now investigate the possibilities for each of these, initially for liquid wastes and then for solid wastes.

4.1.3 Disposal of Liquids to Sea

The most important and controversial liquid discharge from nuclear power stations or fuel cycle plants is that from Sellafield. Here, for the past few decades, low-activity liquid effluent has been discharged via a pipeline. The corresponding authorisation has become progressively more stringent over a period of decades: the main points of the version dated 1.1.85 convert to the following limits (TBqs).

Table 4.1

Sellafield Pipeline Discharge Authorisation (1.1.85)

Period	All Betas	Ruthenium-106	Strontium-90	All Alphas
Any two consecutive days	260	56		1
Any three consecutive months	1,850	370	185	22
Any twelve consecutive months				7

Further, the average number of grammes of tributyl phosphate per cubic metre of discharge should not exceed 300: in any event, best practicable means shall be used to keep the above discharges as low as possible. (It should be noted that the term 'practicable' does not necessarily mean that a reasonable cost has been incurred with respect to the benefit achieved.) Despite the fact that the pipeline discharges have always been well within regulatory limits and an infinitesimal addition to ocean radioactivity (Appendix 1.2), there has been much dispute over the corresponding environmental effects. A complicating feature is that, though the great majority of the radioactivity is swept away within a few days, a small proportion of discharged nuclides can be returned to shore via wind-blown spray or through pickup on sediments eventually deposited on the beach. However, on the whole, such activity attached to solids is transferred away naturally from the Sellafield area over a period of years (RWMAC, 1985). The discharged activity can be tracked through the North Channel of the Irish Sea and northwards off the West Coast of Scotland past Cape Wrath. Statements in the media (Section 9.7) suggesting large accumulations of plutonium near Sellafield therefore have no substance.

An investigation into the above average number of cancer deaths at Seascale near Sellafield concluded that the cause was not necessarily connected with the pipeline discharges (as discussed in Section 6.2.4). Nevertheless, a new treatment plant has been installed to clean up discharges still further; this was followed by an even more stringent and detailed regulation enforced from 1.7.86. Overall, annual discharges have been reduced in general terms about tenfold since the early 1970s. Though these reductions have been greeted as a victory for anti-nuclear bodies and a defeat for BNFL, the latter will have made a satisfactory profit on the new cleanup plants, the costs being eventually borne by the electricity consumer.

It is difficult to see what further means could be used to reduce the effects of such discharges, though one possibility could be the pre-equilibration of discharges in tanks with seawater and sediments, the latter then being collected as sludges and dumped in deep water.

There is a similar regulated discharge by pipeline at Dounreay; perhaps because of its remoteness and lower scale of operation, less controversy has arisen.

4.1.4 The Potential for Disposal of Liquids Under the Sea-bed

The other form of liquid waste, not discharged by pipeline to sea, is HAL at Dounreay and Sellafield (Section 3.2.4). These concentrates are currently stored with the intention to solidify them before disposal. The decay heat of HAL falls off with time in a similar manner to spent fuel or HLW, as shown in Fig. 4.1. An alternative, more direct route to disposal, however, could be to transfer the liquors into a deep borehole near the site, preferably under the adjacent sea-bed, as illustrated in Fig. 4.2. As shown in Appendix 4.2, such a spreading of the heat output from the waste along the long length of the borehole would avoid thermal problems in the surrounding rock. Since HAL is denser than seawater, the system would be physically stable: the introduction of liquor would, of course, be at the bottom of the borehole to displace the seawater upwards before the hole is plugged above the liquor. This form of disposal of toxic liquors is well known, particularly in the US, where natural rock fissures have been additionally useful: so far, there has been no application to radioactive wastes. An aspect of this which would need investigation is the behaviour of possible releases of gases in the liquors or surrounding groundwaters by the radiation from dissolved nuclides.

It would seem difficult, with the present public concern over radioactive waste, to suggest that HAL should be disposed of in the above manner. A more probable waste for this type of disposal, however, could be the currently stored Magnox sludges described in Section 3.2.4. The disposal form could either be a slurry or a solution in acid. Other methods sugested to date are complex and very expensive.

Further discussion on liquid disposal underground is given in Appendix 4.5.

4.1.5 Disposal of Solids On the Sea-bed

Sea dumping has been practised for many years by a number of countries, prominent among them being the

Figure 4.1: HEAT DECAY OF HLW OR SPENT FUEL

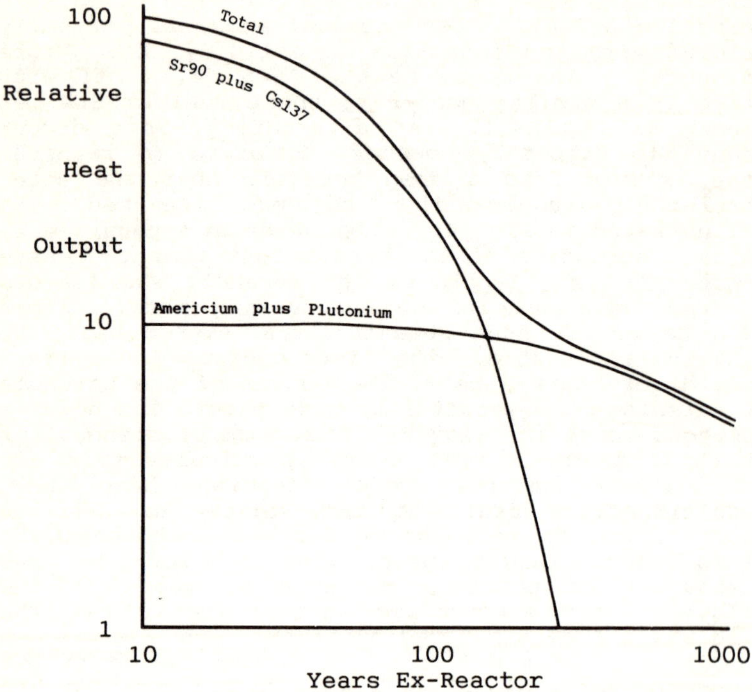

Figure 4.2: POSSIBLE METHODS FOR UNDERGROUND DISPOSAL OF WASTE LIQUORS AT SELLAFIELD (Not to Scale)

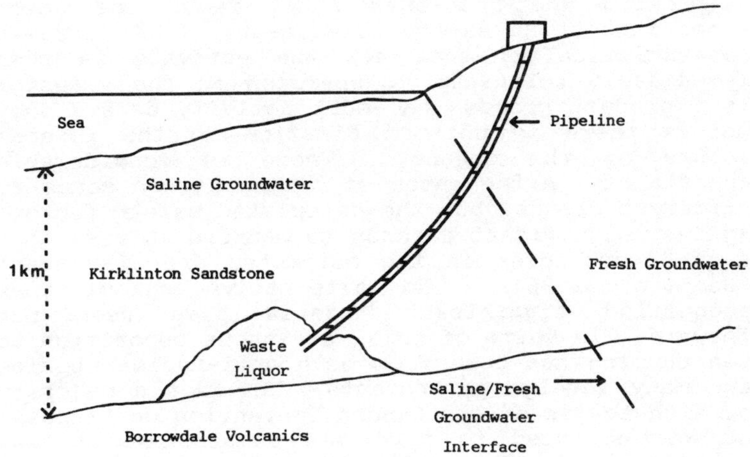

(a) INJECTION INTO PERMEABLE STRATA

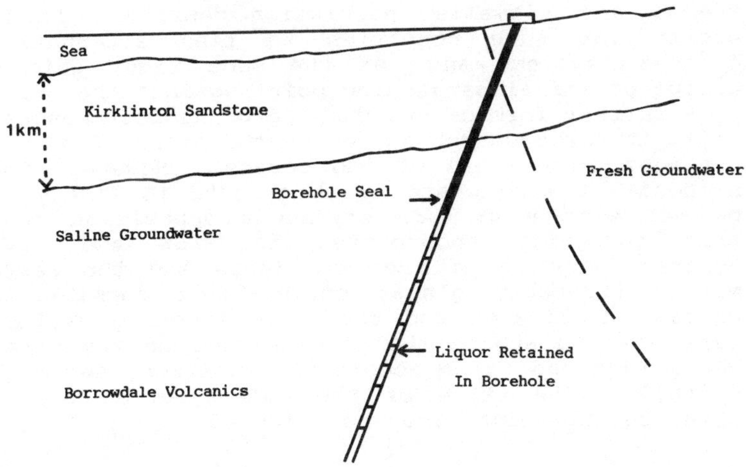

(b) INJECTION INTO IMPERMEABLE STRATA

UK. Under stringent limits for disposed activity set out by the International Atomic Energy Agency (IAEA) and under official supervision at the Atlantic dumping site, drums of LLW and ILW have been dropped overboard (see Fig. 4.3(a)). It is stipulated that packages shall reach the ocean floor intact: in safety assessments, it is assumed (pessimistically) that all the activity is then immediately released. Dispersion in the seawater is a gradual process and much activity decays away before there is uniform dilution in the immense volume of the oceans. There is considerable uncertainty in the rates of mixing betwen zones at different depths, but the calculated safety factors against significant hazards to man are so vast that such inaccuracies in the estimates are irrelevant (Camplin, 1986). No alternative calculations suggesting significant hazards have been put forward. In spite of this, strenuous opposition to sea dumping has come from both anti-nuclear bodies and many foreign governments. In fact a majority of signatories of the London Convention on Disposal of Wastes voted to have sea dumping of nuclear waste stopped. Though the UK continued dumping for a while after the vote, the seamen's union blacked the ships and dumping has ceased, at least temporarily. The above sequence of events is a good illustration of the power that anti-nuclear groups and other interested parties can wield. Many countries have smaller nuclear programmes than the UK and a smaller population density. It is easier for such countries to find alternative disposal systems and, at the same time, gain a useful political bargaining point against the UK.

A further form of sea dumping of nuclear wastes, still in the research stage, is the disposal of HEW packages on the bed of the oceans. However, the disposal of such waste by this route is likely to be met with even more strenuous opposition than that currently encountered for ILW and LLW. Further, rupture of the containers and the waste matrix (probably glass) could be a possibility unless complicated and expensive lowering systems were used to ensure the package reached the ocean bed intact and in an acceptable position (see Fig. 4.3(b)). The value of the matrix as a barrier after rupture might then be disputed.

4.1.6 Disposal of Solids Under the Sea-bed

Two basic methods of emplacing nuclear waste packages under the sea-bed are by free fall with penetration on impact (Ove Arup, 1985) or by drilling and emplacement in the borehole (Bury, 1985) (See Fig. 4.4). After putting a seal over the package where necessary, any escaping activity would have to pass through

(a) the material of the waste matrix and the container, and

(b) the rocks and/or sediments around the hole, assuming the sealing procedure was as good an inhibitor to nuclide migration as its surroundings.

Finally, such activity which did reach the sea would be subject to the delay of dispersion and enormous dilution described in the on-seabed option above. Further, for the saline groundwater conditions round a package, laboratory investigations have shown that migrating nuclides are picked up by minerals in the rocks and/or sediments so that their rate of movement is far slower than the groundwater. On the other hand, there is an upwards convective effect caused by the heat given off by the waste, rather like the circulation induced by a boiler in an unpumped household central heating system.

One hardly needs to carry out sophisticated calculations to be convinced of the adequacy of safety after the package has been satisfactorily sealed beneath the ocean bed. There is a considerable effort to improve basic data and designs, e.g. at the Woods Hole centre of oceanographic expertise in the US and the European Research Centre at Ispra in Italy (Freeman, 1984 and Murray, 1986 and 1987). However, some aspects pose difficulties.

A. As in Section 4.1.5, technical doubts exist in the free-fall method in being certain that there are no hard obstacles in the sea-bed, causing possible rupturing of the packages and sediment cover.

B. In disposal at the great depths of the ocean bed, the operations of excavation, emplacement and sealing are difficult, and their reliability and the feasibility of remedial action after a fault are uncertain. These difficulties still occur, though somewhat reduced, when drilling is done in shallow coastal waters, as in the 'ENSEC' approach (Richards, 1985), where a jackup platform, commonly used for drilling for oil beneath the sea-bed, replaces the ship, in Fig. 4.4(b). (This is effectively the USO method defined in Section 4.1.8.1)

C. There could still be international opposition to ocean bed disposal and perhaps also in offshore waters, where regional opposition groups would no doubt be organised.

The first two difficulties are assessed as unimportant by Ove Arup (1985); the chief constraint is likely to be the last one above, i.e. international opposition.

4.1.7 Disposal of Solids by Shallow Land Burial

Much of LLW to date has been buried in shallow pits, eventually covered with local soils. The depth of cover has usually been defined by regulatory authorities to allow a limited time of access for inspection of the surface; this period, because of the short half-lives in the waste, would probably become unlimited by the time the site was vacated. In the UK, most LLW has been deposited at Drigg in Cumbria, in trenches based on glacial clay, which overlies sandstone. Drainage from the clay bottom collects in a small stream, which itself discharges into the Ravenglass Estuary. Doses at the dump surface and concentrations of alpha and beta nuclides during disposal are subject to regulation. Frequent monitoring has revealed negligible atmospheric release of activity; stream sampling shows that an individual would have to use the stream more or less continually as a sole supply of drinking water to receive a dose that would approach the ingestion limit recommended for the public. In fact, only cattle use the stream for drinking: details of nuclides found therein are listed in BNFL (1986).

Figure 4.3: DISPOSAL OF WASTE PACKAGES ON THE OCEAN BED (Not to Scale)

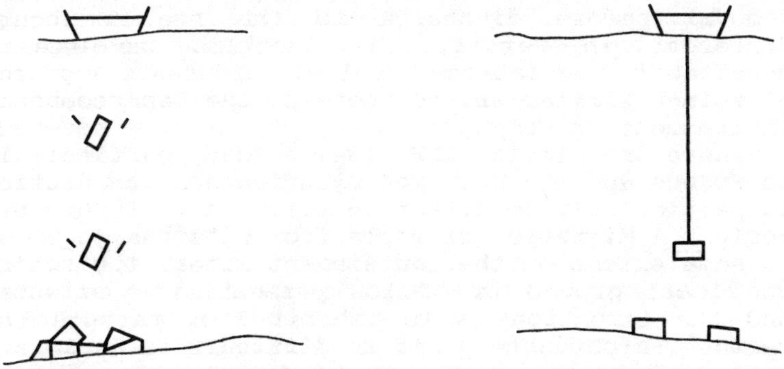

(a) OVERBOARD DUMPING (b) LOWERING

Figure 4.4: DISPOSAL OF WASTE PACKAGES UNDER THE OCEAN BED (Not to Scale)

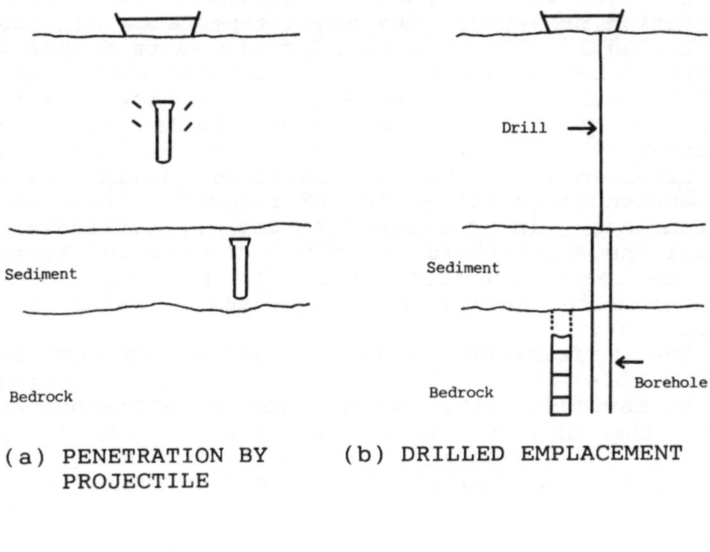

(a) PENETRATION BY PROJECTILE (b) DRILLED EMPLACEMENT

The other UK site for LLW disposal is at Dounreay in Caithness. Here, pits have been dug in hard rock. Drainage from emplaced waste is pumped out to the site flocculation plant for solids removal before discharge to the sea. Though different in detail, the Scottish Development Department regulations control disposals to the pits in a similar way to those of the Department of Environment at Drigg.

There are similar LLW sites abroad, particularly in France and the US. The experience of the latter is particularly important in relation to future pit designs. Migration of waste from pits has occurred to some extent on the containment sites, i.e. where the local ground is of low permeability to water and the intention is to inhibit flow through the waste. Unfortunately, it is difficult to explore a site so comprehensively as to detect all zones of higher permeability. Further, there can be sufficient retention of water by surrounding ground as to cause flooding of the pit. The waste is then subject to continual submersion and leaching, with a throughflow of water, through thin permeable sections or overflows, allowing migration. This can occur through lateral flows even where an impermeable cap is installed. Remedial action, once activity has started to move, is difficult. Such a situation is commonly known as the 'bathtub' effect (see Fig. 4.5). On the other hand, US nuclear waste 'drained' sites, where the bottom of the waste is above the water table at all times (see Fig. 4.6), have not revealed significant migration problems. The above experience has been reflected in US regulations for LLW waste disposal, where it has been expressly stated that containment sites should not be considered unless locations for emplacement above the water table are not available.

Experience in non-radioactive toxic waste management disposal in the UK reveals similar poor performance in containment sites, particularly where the water table is between the top and bottom of the waste and fluctuates. On the other hand, drained sites have behaved much more satisfactorily (DOE, 1978).

The performance to date at Drigg and Dounreay shows how even the simplest drained burial, illustrated in Fig. 4.6(a), can be satisfactory; with the addition of an impermeable 'cap' (Fig. 4.6(b)), new drained pits should be even better. An attractive feature here is that percolation through a crack which conceivably might develop in the cap, would only contact a limited amount of waste, so that any activity picked up would be several orders of magnitude less than at 'capless' pits. A further advantage of the drained type of site is that there is considerable flexibility in location, so that a coastal site draining to sea and near the source of the waste to reduce transport could be chosen.

In view of the above experience, it is surprising that the candidate sites, chosen for investigation in 1986-7 by NIREX for a new LLW disposal facility in the UK to supplement Drigg, were all of the containment type. (In fact, the borehole evidence from nearby areas suggested that the clay at one of the sites, Elstow, was laminated and underlain by permeable strata (Blowers, 1987).) However, these sites were abandoned before the General Election in May 1987; coincidentally, they were all within the boundaries of Conservative-held seats. Clearly, the Conservative Party felt that the hazard of losing parliamentary seats outweighed the alleged safety features of the selected sites! It is to be hoped that any future UK shallow burial sites will be of the drained type.

A rather sophisticated concept of drained trench burial is illustrated in Fig. 4.7, designated for reference as a 'Dry Box'. Here, radioactive waste is enclosed in a concrete 'box', vented near the top to release gases from decomposing wastes such as plastics and drained at the bottom to prevent any accumulation of liquids. The intrusion barrier required by regulatory authorities is composed of two parts, the upper and lower roofs, constructed in concrete. Access is then possible between the roofs for inspection and remedial action to ensure that no water can drain into the waste. With the ability to inspect and seal any cracks that do appear, particularly in the underside of the upper roof, the concrete will remain an effective barrier for many centuries. Extra safety can be provided by an impermeable layer such as clay on top of the upper roof, which is itself supported on concrete pillars through which drainage can be arranged. Land drains in soil or rubble above the impermeable layer can lead away the preponderance of percolating water. Inspection can be arranged for the underside of the box if desired. In fact, the system has the attraction of being <u>seen</u> to be satisfactory, with the possibility of remedial action, checkups becoming less and less frequent as the radioactivity decays away. It might indeed be described as a 'Rolls Royce Convertible' design, gradually changing over from storage to disposal! The system is flexible; for example, Dry Box units could be set out linearly as part of sea defences, behind which conventional non-active waste could be used as landfill in (say) the reclamation of coastal marshes. French mounded concrete bunkers and Canadian shallow below-ground vaults have some similarities to the Dry Box concept. A variation which might be attractive in the future could be to dispose LLW in concrete pressure vessels remaining after decommissioning reactors to Stage 3 (Section 3.2.5). After earthing over, such disposal units could be less objectionable aesthetically or toxically than dumps of coal wastes.

The activity of the majority (by volume) of decommissioning wastes is in low-activity concrete or in induced activity in materials of low leachability such as stainless steel; submersion of these materials in water should therefore not have serious consequences. Two concepts which might be considered here are illustrated in Fig. 4.8. In the example of Fig. 4.8(a), decommissioning waste or rubble could be used to raise the foreshore to permit drained disposal of LLW or decommissioning wastes, finally sealing with an impermeable cap. In the example of Fig. 4.8(b), the foreshore has been excavated using a coffer dam. After burial of waste at a level below that of low tide, with coverage by nuclide sorption material as desired, the sea and sand are allowed to flow over the emplacement zone. The inherently flat surface of the sea ensures that there are no significant differences laterally of water pressure to induce water movement near the waste.

A variation of the latter example involves the dredging of sea-bed trenches in shallow waters, emplacing waste therein and covering it with subsequently dredged material; the whole sequence of operations could be done from a specially designed dredger. This variation could be especially useful for disposing large items of low-activity waste, such as heat exchangers from reactor decommissioning.

Figure 4.5: TRENCH BURIAL BELOW THE WATER TABLE ILLUSTRATING THE 'BATHTUB' EFFECT

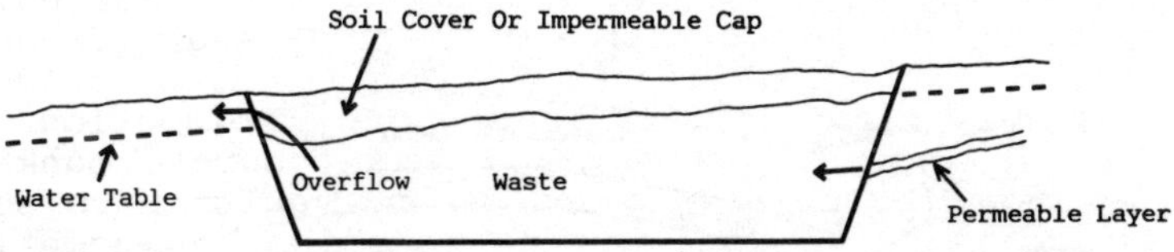

Figure 4.6: TRENCH BURIAL ABOVE THE WATER TABLE

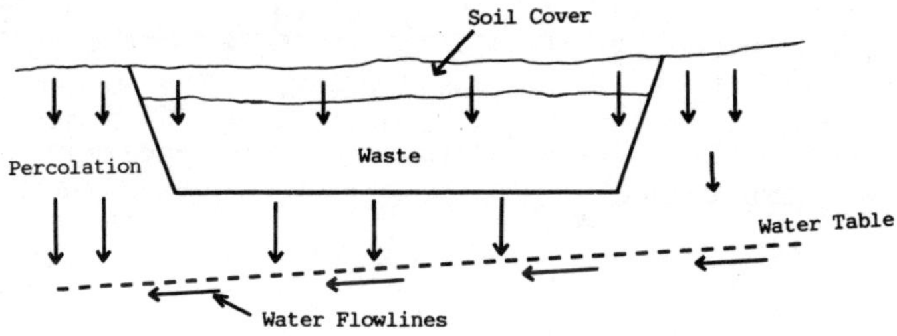

(a) SIMPLE SOIL COVER

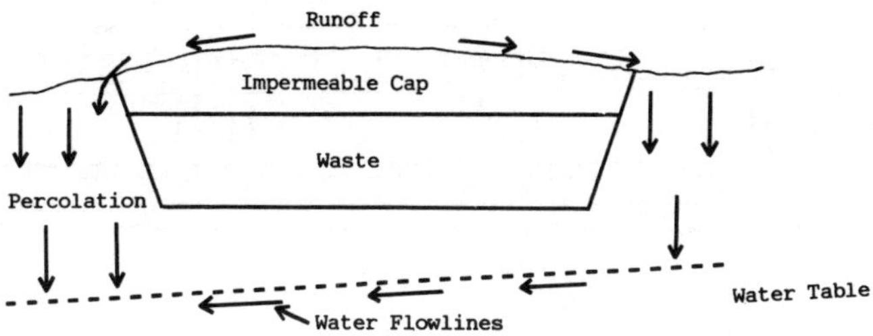

(b) IMPERMEABLE CAP

Figure 4.7: 'DRY BOX' CONCEPT FOR SHALLOW LAND BURIAL OF RADIOACTIVE WASTE

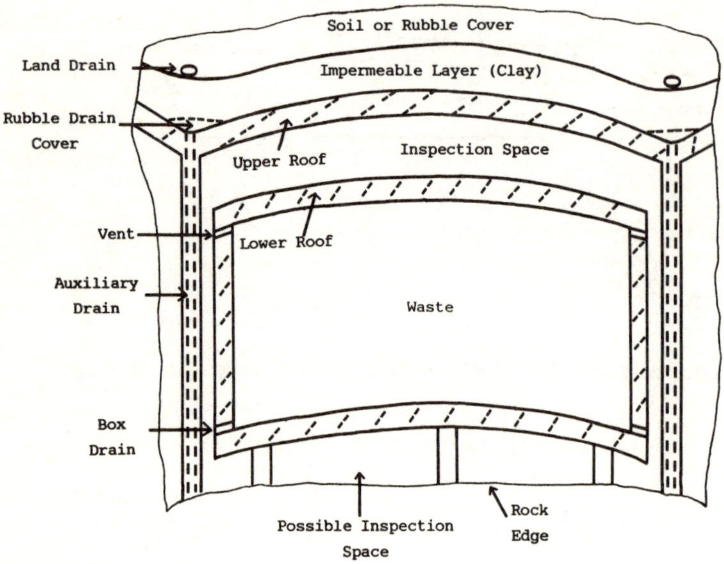

(a) END VIEW

(b) LINEAR LAYOUT OF DRY BOXES - PLAN VIEW

Figure 4.8: WASTE EMPLACED AT THE FORESHORE

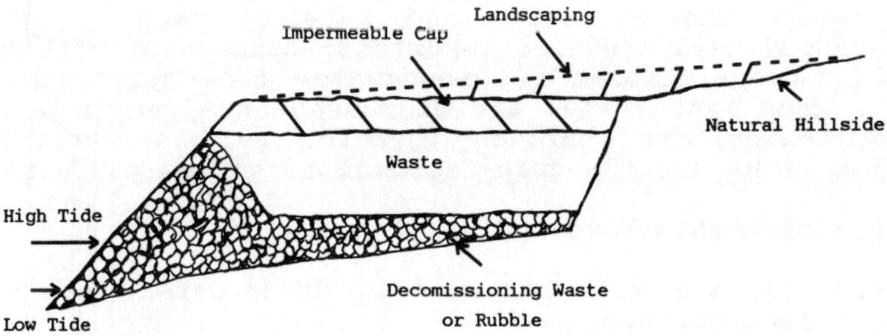

(a) WASTE EMPLACED ABOVE THE TIDAL RANGE

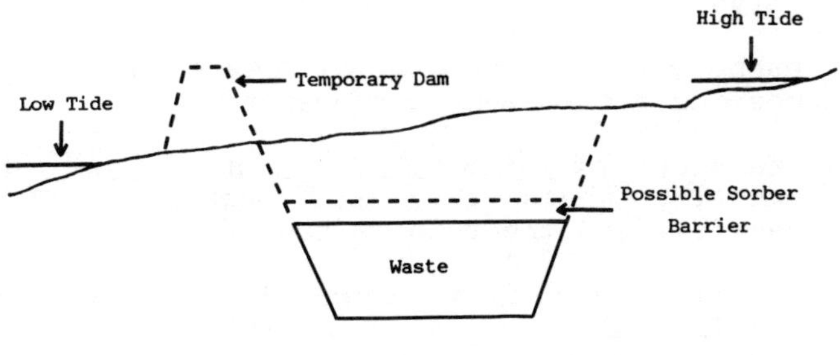

(b) WASTE EMPLACED BELOW THE TIDAL RANGE

4.1.8 Disposal of Solids by Deep Land Burial

4.1.8.1 General. For nuclear wastes containing long-lived activity (HEW, Spent Fuel, PCM and some forms of ILW) burial underground must be deep, with the aim of providing a long migration path to the biosphere along which many nuclides will decay away. The direction and rate of flow of the groundwaters are also important; consequently, some of the principles of groundwater flow appropriate to deep land burial are discussed in Appendix 4.3. Following the approach therein, we can classify locations for the deep disposal of solids as below.

(a) Disposal above the water table,

(b) disposal with surrounding groundwater flow to land surface,

(c) disposal in saline groundwater, and

(d) disposal in fresh groundwater discharging to sea.

It is useful for reference hereafter to define the three deep burial systems currently under consideration by NIREX (UK NIREX, 1987):

(a) burial under land with access from land: this we denote as ULL,

(b) burial under the sea-bed with access from land: this we denote as USL, and

(c) burial under the sea-bed with access from offshore: this we denote as USO.

For each type of location, there are basically two engineering emplacement arrangements:

(a) transfer of packages to positions inside tunnels, where they are packed round with backfill and the tunnels sealed, and

(b) lowering packages down relatively close-fitting boreholes with backfilling round and between packages carried out at considerable distance from the operators.

A further consideration arises when packages emit a significant amount of heat. Rocks in general are poor conductors and high temperatures can be generated unless appropriate geometrical configurations are laid down. The basis of such heat transfer is discussed in Appendix 4.2. We now turn to the location types listed above.

4.1.8.2 Disposal above the Water Table. Figure 4.9 illustrates a typical disposal system above the groundwater table (a more detailed drawing of this 'Dry Repository' is given in Burton (1981), Fig. 2). Before excavation begins, the water flows are in principle of the form outlined in Appendix 4.3 and shown in Fig. 4.10. Emplacement tunnels are excavated above the water table sufficiently far from the weathered zone of relatively high permeability (see Fig. 4.9) to give only a very slow percolation in the tunnels through any tiny fissures that occur. The access tunnel is then arranged to provide drainage, with branch tunnels and boreholes where necessary, so as to bring the water table down below the emplacement tunnels. Observation of the latter when empty will reveal any seepages which can be sealed and/or avoided when emplacing the packages.

One way of emplacement is shown in Fig. 4.9(c) with packages resting on a granular bed in a tunnel, with further material added to fill the tunnel. The granular material could be chosen for its ability to pick up key nuclides if leached by water. The design is discussed in detail (Burton, 1981) for a package of HEW surrounded by a shield thick enough to allow 40-hour per week access after 100 years and for shorter periods before then. Surface storage of these shielded packages for 100 years can reduce tunnel excavation requirements, which are dependent on heat output (see Appendices 4.1 and 4.2). Alternatively, a shaft connecting emplacement tunnels to the surface above can act as a chimney when packages are emplaced during this period, inducing an air draught past the packages.

There are several advantages to this type of disposal system.

(a) The tunnel can be left without any backfill for as long as desired, during which time the integrity of the tunnel walls to seepage can be seen to be satisfactory.

(b) There are a number of coastal hillsides in the UK of satisfactory rock and of sufficient height above sea-level both to avoid flooding of the wastes during possible melting of the Earth's ice caps and also loss of cover through erosion, at least for many thousands of years.

(c) Little wetting of packages can occur both because of the thick shield and because most water will bypass them: if leaching somehow did occur, the granular backfill would inhibit any nuclide migration. Again, if nuclides passed through backfill, the environment impact would be infinitesimal, especially at a coastal site, where any contaminated drainage would be enormously diluted in the sea.

Among obvious variants, the package could be unshielded. This would save on shielding costs but require remote operations for transfer, emplacement, inspection and backfilling the tunnel; one barrier, moreover, would be lost. As a further alternative, unshielded packages could be lowered down boreholes in the tunnel floors. However, there would have to be drainage from the boreholes into lower auxiliary tunnels: inspection and backfilling could also be difficult. Furthermore, if the packages were heat emitting, the boreholes would then have to be spaced so far apart to avoid thermal interaction that the same length of tunnels would be required as in the basic tunnel emplacement system (See Appendix 4.2).

The above discussion and safety assessment, detailed in Burton (1981), show that a Dry Repository as above can be conceived for HEW: a number of locations exist in the UK with an acceptable topography. The wastes with lower heat outputs, PCM and ILW, clearly could also be satisfactorily disposed of in a Dry Repository. However, in the case of spent fuel, the safety would hinge on the migration of plutonium, which would be present to a level of about one hundred times greater than in HEW. On the other hand, retrievability would be easier to design than for the Wet Repository designs described later, so that long-term storage, with the flexibility for either retrieval or final sealing up, could be an attractive option possible with a Dry Repository.

4.1.8.3 <u>Disposal Below the Water Table with Groundwater Flow to the Land Surface (ULL)</u>. This type of disposal has been investigated by several countries, especially the US, where the distances from the waste sources to the coast are considerable. It is clear from Appendix 4.3 that, of the rain falling on land, the small fraction which percolates through the ground must create sloping groundwater tables in order to cause such percolation to move to lower surface levels or the sea. This slope affects deep groundwater, the consequent variations of pressure heads inducing flow in the region of a repository (see Fig. 4.10). It is difficult to forecast the eventual path of all water passing through the repository, both in length to the surface and the nuclide pick-up properties of the rocks en route. Together with the rather arbitrary nature of the behaviour after discharge to the surface, mentioned earlier, the prediction of dosage to man in drinking water, though probably extremely low, is uncertain.

Since the waste is immersed in water after cessation of supervision, this concept has been called a Wet Repository (Burton, 1981). A typical layout is illustrated in Fig. 4.11 with access via adit (sloping tunnel). (According to the characteristics of a particular site, either shaft or adit access may be preferred.) Emplacement of either shielded or unshielded packages in tunnels could have some similarities to that of a Dry Repository, but the backfill round a package would be of much smaller particle size, such as clay, in order to inhibit water movement near the packages. In effect, this is a form of containment disposal discussed for shallow land burial. The intention would be to cause any flow in cracks in adjacent rocks to bypass the tunnels. Emplacement in boreholes in tunnel floors would need no connection to auxiliary drains: however, backfilling round packages to create a substantial barrier to water flow would be more difficult than where the emplacement is in tunnels.

Locations for a Wet Repository, at greater depths than available in the UK for a Dry Repository, would not be difficult to find, but the hazards of construction increase with depth, because the large water heads may at any time induce rapid inflow through a fault as the tunnel face is extended. The establishment of safe disposal provides the same dilemma as in other containment systems, that the more extensive the borehole tests to verify the absence of faults in the rock, the greater the likelihood that a bore-hole develops a fault in its sealing in the future.

An example of a comprehensive study into deep waste burial is that carried out by the KBS division of the Swedish Nuclear Fuel Supply Company (KBS, 1983). The objective was to show that spent fuel could be disposed safely by deep land burial in Sweden, since under Swedish law such an objective must be achieved before fuel can be charged into a reactor. The designers suggested the following stages leading up to final disposal.

(a) Storage of spent fuel under water for 40 years.

(b) Encapsulation of the fuel in a 100-millimetre thick copper container with the internal voids filled with either copper powder or lead.

(c) Emplacement of the resulting packages, each containing one sub-assembly of fuel, in individual holes in the floors of tunnels 500 metres deep. The spaces around the packages would be filled with highly compacted bentonite, a form of clay which swells on contact with water and is highly retentive of migrating nuclides.

(d) Final backfill of the tunnels with a mixture of sand and bentonite.

A very large research programme on all features suggests the following safety characteristics of the above procedure.

(a) The copper container and bentonite surround will prevent water access to the fuel for millions of years.

(b) The eventual leaching of the fuel will be so slow because of the solubility effects that there will be no significant increase in the natural radioactivity of the groundwater.

(c) Inadvertent intrusion into the repository is inconceivable.

(d) Low-frequency events such as earthquakes are unlikely to cause serious damage.

The overall conclusion of the project was that such a disposal scheme was very safe and that economies could be made in several areas. This latter point is not surprising, considering the large requirement of container material (for each tonne of fuel, either 8 tonnes of copper or 5 tonnes of copper and lead are required!). Moreover, the disposal locations were presumed to be in fresh groundwater migrating to drinking water supplies: inherently much safer locations would occur in saline groundwater. It should be noted, however, that the study applied to oxide fuels with relatively unreactive cladding. In the case of Magnox fuels, both the uranium metal fuel and the magnox alloy cladding can undergo reaction with groundwater, releasing considerable amounts of gases and chemical energy as heat. This could considerably affect safety assessments if disposal of this kind of fuel were proposed.

An obvious alternative to using access by tunnel to disposal zones is to drill very long boreholes from the surface. This is safer for operators during construction and can indeed provide great depths for emplacement. For example, a borehole 2 kilometres deep could be drilled. However, in order to provide 1 kilometre of cover for the top package, only the bottom half of the hole would be filled with waste. This system would seem not so attractive for larger volume wastes, such as PCM and ILW, which could be packed efficiently into tunnels without heat problems. On the other hand, for smaller volumes of wastes with relatively high concentrations of long-lived nuclides, such as spent fuel or wastes containing iodine-129 (half-life 16 million years), the great depths to which modern equipment can drill might provide an attractive method of disposal. Further details on the practicality of drilling very deep holes of various diameters is given in Appendix 4.6.

Some notional costs for Dry or Wet Repositories (Burton, 1981 and Griffin, 1982) and for very deep borehole systems (Appendix 4.6) show that the costs of disposal of the relatively small volumes of nuclear wastes other than LLW are likely to be an insignificant proportion of the value of the electricity corresponding to the waste. The larger volumes of LLW make deep burial in its case less attractive.

Before leaving the discussion of the above designs, it is perhaps worth mentioning the subject of monitoring. It is often claimed that a disposal site will be monitored after sealing. It is rarely said, however, what action would be taken if radioactivity were found to be migrating. People living near a proposed disposal site often ask for an assurance that the waste will be removed if faults develop. However, operations in radioactive ground conditions can be awkward and expensive. It is far better to include in the design the ability to take adequate remedial action, without necessarily removing the waste. For example, grout might be pumped into developing rock cracks before significant spread of activity occurred. In respect of such early warning monitoring and effective remedial action, dry disposal in a deep repository or drained shallow burial has considerable advantages over wet disposal under the groundwater table.

4.1.8.4 Disposal in Saline Groundwater. Appendix 4.3 outlines the principles of groundwater flow, in particular near the coast, as for USL (Fig. 4.10). Clearly, if waste can be emplaced sufficiently seaward from the saline/freshwater interface, two desirable safety features are obtained as mentioned previously, i.e. there are no differences of pressure to induce water movement near the waste and there is a vast dilution in the sea of any activity which escapes from the sea-bed. This dilution will probably be more immediate and effective than for the normal pipeline discharges, where some nuclides are picked up on sediments and returned to shore (Section 4.1.3). This occurs because any of the latter types of nuclides which might migrate from a repository through the rocks of the sea-bed are likely to be picked up by minerals in the rock before emergence, since many of these minerals will be common to those in the local sediments.

The above features can compensate against choosing some inland site where the rock has (say) a lower permeability or a superior capability to pick up migrating nuclides. A further attraction is the possibility of straight transfer from waste sources on the coast, in particular at Dounreay and Sellafield, without the need for special packaging to satisfy regulations for transport through the public domain. The overall advantages above induced the author to propose investigations into constructing a repository under the sea-bed off Sellafield some ten years ago (Burton, Internal BNFL Document, 1979). A later report (Griffin, 1982) showed that appreciable cost savings could be gained by not transporting the waste for disposal over long distances. It is interesting that the intention to look into such a possibility has at last been declared by BNFL (in September, 1987). Even if the bed of the Irish Sea should, at some time within the relevant future, become dry land, as some sources have predicted, the groundwaters below the existing sea-bed can remain effectively stagnant and saline for millions of years, decaying away radioactivity in waste to insignificant levels. At Dounreay, it is unlikely that the adjacent sea-bed will become dry land within the effective existence of disposed activity.

Fig. 4.11 gives an elevation view showing access from land by an adit to a Wet Repository under the sea-bed (a corresponding access by shaft is given in Burton (1982, Fig.3)). A U-shaped adit has been indicated, since this can inhibit convective water movements in the (remote) possibility that the adit develops a fault after sealing. Details of a possible system of emplacement tunnels at the end of the adit are shown in Fig. 5 of Griffin (1982). Options for emplacement within tunnels in this case are similar to those described in Section 4.1.8.3; the safety case after disposal at the same depth, however, is superior, for reasons indicated in the previous paragraphs.

A disadvantage of the USL concept is that the necessary exploration to establish an acceptable location could be expensive because of the extra difficulties of drilling test boreholes offshore. Costs would be expected to be considerably greater than for drilling from a land surface, as listed in Appendix 4.6.

A natural extension of this system is to site it under sandbanks. These are often a few kilometres away from the coast, i.e. giving sparse local population, yet still in UK territorial waters. An attractive example is the system of sandbanks in the Wash, which overlay considerable thicknesses of clay strata (see Figs 10.8 and 10.9); the cost of constructing plants on 'fill'-type islands on sandbanks is, if course, much less than those in deep water requiring extensive foundations (Binnie, 1982). If waste emplacement took place under a few hundred metres of clay beneath sandbanks or deeper seawater, as illustrated in Fig. 4.12(c), it is estimated in Appendix 4.4 that even plutonium from spent fuel would never reach the sea. Because of the relative difficulties of tunnelling in clay, boreholes drilled from the surface might be preferred for emplacement of wastes.

Slant drilling of boreholes from land to regions of saline groundwater are yet another possibility, as shown for liquids in Fig. 4.2.

Migration of nuclides induced by convection caused by heat emitted by the wastes must be evaluated (see Fig. 4.12(a)). One possibility to restrain such convection is to locate the waste in or under soluble salt deposits or layers of high salinity groundwater. The decrease in groundwater density caused by the heat from the waste may then be insufficient to make the deeper groundwater less dense than that above. Convection will therefore cease at the interface (see Fig. 4.12(b)). Such zones of salts or high salinity exist in several regions both under land and the coastal sea-bed of the UK.

4.1.8.5 Disposal in Freshwater Discharging to Sea. This type of situation is intermediate between systems of Sections 4.1.8.3 and 4.1.8.4. Possible locations are shown diagrammatically in Fig. 4.10(b) with a hydrological discussion in Appendix 4.3. Basically the repository is under land but in a flow of groundwater which does not emerge above sea level and also is unlikely to be used as drinking water. According to the disposition of local strata, discharge can occur at the foreshore or be carried out to an appreciable distance under the sea-bed.

Emplacement operations are similar to those of the Wet Repository concepts described above. The safety after disposal, however, is intermediate between those of Sections 4.1.8.3 and 4.1.8.4, in that, although there is the advantage of dilution of any escaping activity in the sea, the groundwater heads inland can induce significant flows in the region of the repository and direct migration of activity along a relatively short distance to the biosphere. Thus, apart from some saving in exploration costs, there would appear unlikely to be as good safety characteristics in this type of location compared with nearby situations in saline groundwater under the sea-bed. With the availability of slant drilling, this advantage would also seem to be true for any borehole system from the land surface to saline groundwater under the sea-bed, rather than to fresh groundwater under land.

Figure 4.9: TYPICAL LAYOUT OF A DRY REPOSITORY
(Not to Scale)

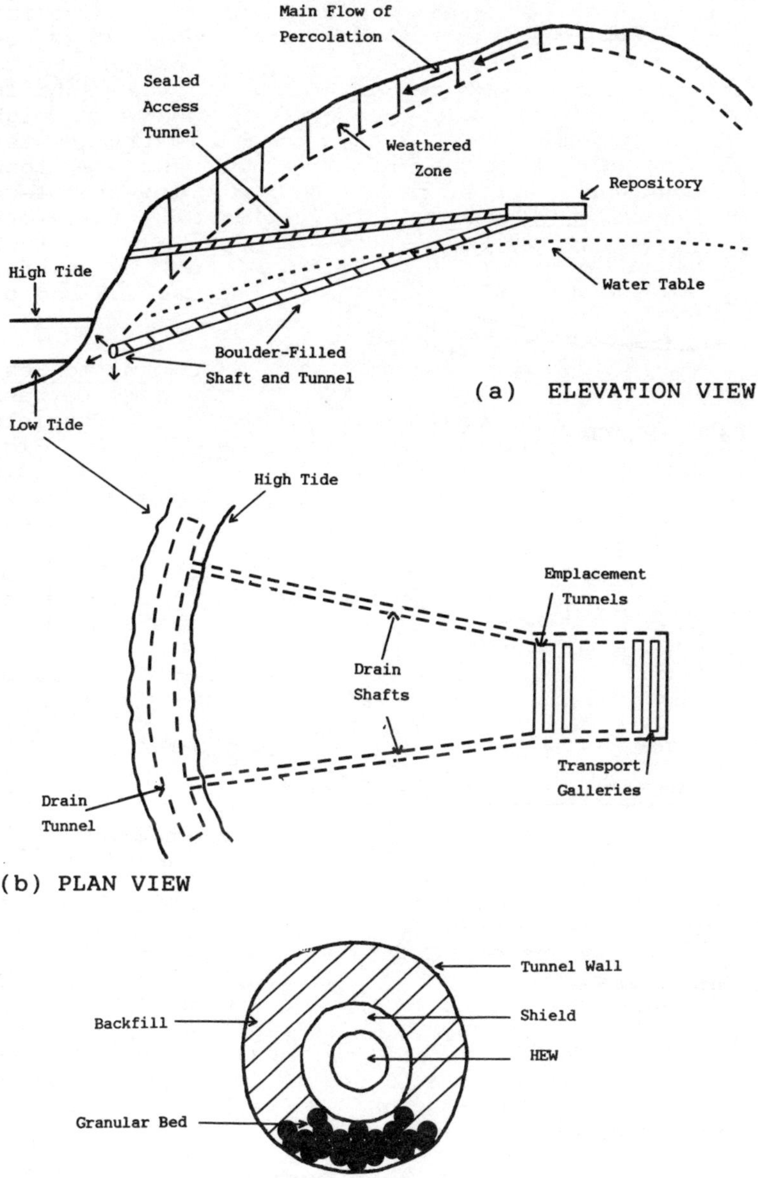

(c) END VIEW OF EMPLACED PACKAGE

Figure 4.10: OUTLINES OF GROUNDWATER FLOW

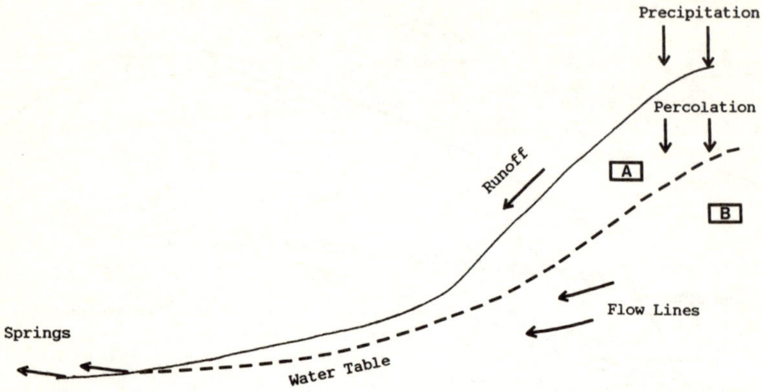

(a) INLAND

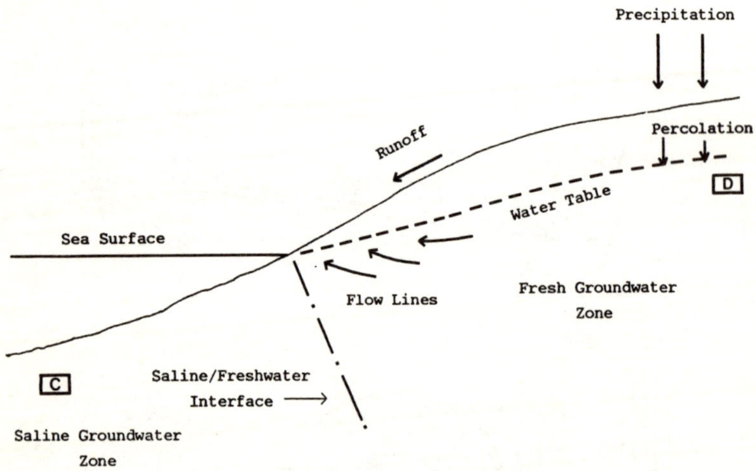

(b) AT THE COAST

Note: A, B, C and D are possible general repository locations discussed in Section 4.1.8.

Figure 4.11: USL FORM OF WET REPOSITORY
(Not to Scale)

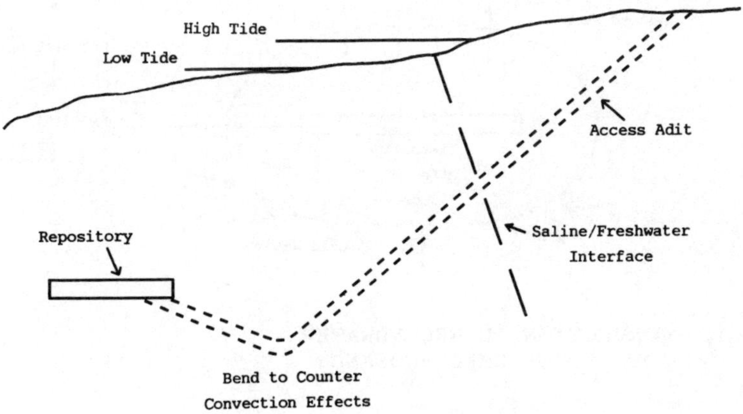

(a) ELEVATION VIEW

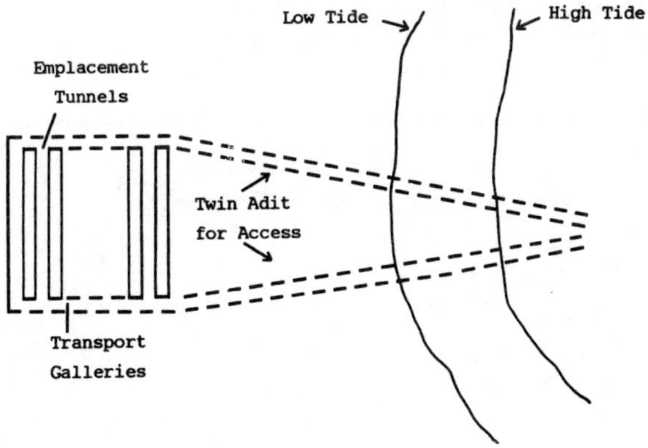

(b) PLAN VIEW

Figure 4.12: MIGRATION OF NUCLIDES THROUGH CONVECTION AND DIFFUSION

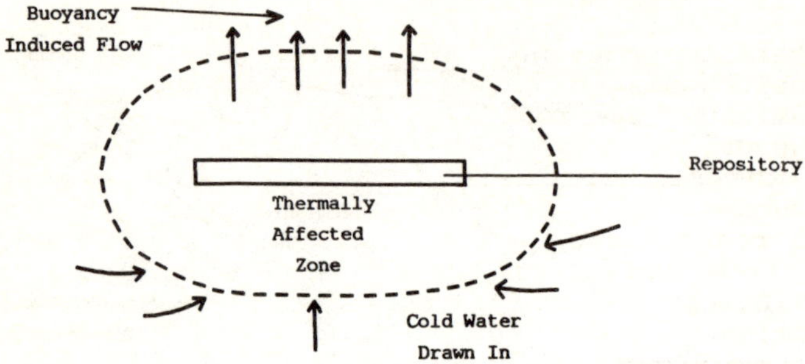

(a) CONVECTION IN GROUNDWATER DUE TO THERMAL BUOYANCY

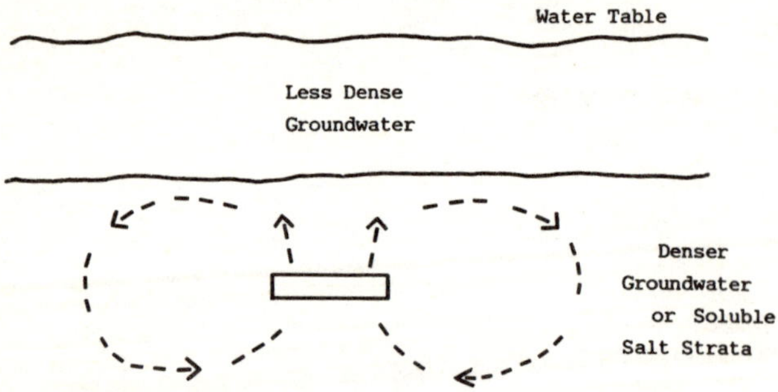

(b) STRATIFICATION INHIBITING CONVECTION TO THE BIOSPHERE

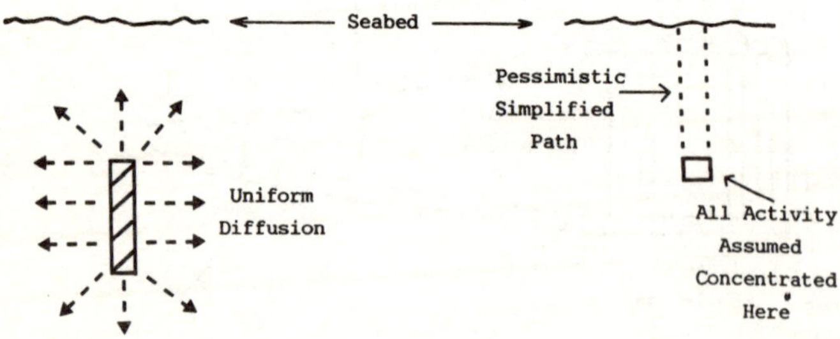

(c) DIFFUSION MIGRATION OF NUCLIDES WITH CALCULATIONAL SIMPLIFICATION

4.2 COAL-FIRED POWER

4.2.1 Disposal of Liquids

Water percolating into mines can pick up a considerable quantity of elements such as iron and aluminium, due to its acidity. There is no option, however, in view of the large volumes involved, but to dispose of these to local surface flows or groundwaters. Oxidation on reaching the surface and reduction of acidity can throw down hydroxides of iron and aluminium etc., causing considerable pollution, as in the River Girvan, described in Section 6.3.4. In addition to metal pickup, natural groundwaters can be highly saline, again a source of pollution. Wash water from coal cleaning will probably be of similar quality to mine waters and so is a potential source of pollution. Unlike nuclear discharges, which are regulated by the rates of discharge of radioactivity, liquid effluents from the coal industry are regulated by concentration so that polluting chemicals at low concentration may still have an appreciable output, because of the large volumes of liquid discharged.

4.2.2 Disposal of Solids

There are two main types of solid waste from coal-fired power systems, spoil from mining and washing coal, and the ash left after burning the coal. The first of these has a considerable volume - on average about half the volume of the usable coal. This fraction has increased dramatically since mechanised extraction began, partly since such techniques make backstowing of spoil underground more difficult. Land use has also increased after the Aberfan disaster (discussed later in Section 6.4.2), since spoil slopes have to be made less steep. There are few cases where there is an alternative to tipping spoil locally; as a consequence, nearly 2 million tonnes are dumped per GW(E)yr, requiring 0.1 square kilometres of new land (Comm. En. Env., 1981). The fly ash from the combustion of the coal amounts to about 300,000 tonnes per GW(E)yr, of which some 40% finds commercial use. The rest is dumped in heaps similar to the mining spoil or pumped as slurry to settle in lagoons. This dumped residue requires a further 0.3 square kilometres (per GW(E)yr) of new land adjacent to the power station.

Both heaps of spoil and dumped fly ash contain soluble toxic components which can be leached out by rainwater. For example, 2% of fly ash is soluble and 1% of this is potentially harmful (Comm. En. Env., 1981), containing such elements as selenium, mercury, vanadium and radioactive radium, and thorium isotopes. These toxicants therefore make their way to local streams and groundwaters.

Future desulphurisation processes (Section 1.3) will produce huge quantities of by-products, such as calcium sulphate, which will cause severe environmental problems (Longhurst, 1987).

4.3 FUSION POWER AND RENEWABLE SYSTEMS

4.3.1 Disposal of Liquids

Fusion and renewable processes do not involve liquids directly, though in some cases there can be a connection with the wastes of other processes if these are used in the fabrication or construction stage. For example, the cement needed to construct dams for tidal power will require the combustion of a considerable amount of fuel to heat limestone for making cement. There may also be a small amount of waste liquors from fusion power if lithium is processed for recycle or discharged.

4.3.2 Disposal of Solids

Here, most processes would only require conventional disposal of demolition waste. Again, the exception is the fusion reactor, whose structural material after neutron bombardment could pose similar problems to the steel and concrete which has been close to the core in a fission reactor. A similar disposal strategy might be chosen, i.e. to wait for cobalt-60 to decay away, before final disposal of both operational and decommissioning wastes. The disposal of a tidal barrage, like that of nuclear reactors decommissioned to Stage 3, may never be carried out.

4.4 SUMMARY

The discharge of low-active liquid nuclear wastes by pipeline to sea is carried out under stringent regulations, causing negligible hazard locally and adding an infinitesimal increment to the radioactivity of the world's oceans. Political opposition is nevertheless strong, particularly with respect to Sellafield discharges. The potential exists in principle to drain away higher-active liquids deep underground.

Because of the low volumes involved, the disposal of solid nuclear wastes is technically feasible at locations distant from their source, in particular onto or under the sea-bed. However, international opposition makes the latter disposal routes less attractive than land burial in the UK.

Shallow land burial of LLW has been practised for many years by several countries. Difficulties have occurred when the waste is partially or wholly below the groundwater table. On the other hand, satisfactory experience has been obtained when the waste is above the groundwater table. In spite of this, political opposition is calling for more exotic designs in future. However, there has been no call to dig up waste already buried. If it is satisfactory to leave this untouched, why build more expensive and complicated new facilities? DOE requirements for the protection of the public are now extremely stringent and have been interpreted by NIREX as applying almost indefinitely into the future, i.e. beyond several Ice Ages. Several new concepts can be envisaged for LLW and decommissioning waste burial in trenches which could well ease public anxieties.

Deep land burial of spent fuel and HEW in tunnels or boreholes can be classed according to their position of emplacement with respect to the groundwater table and the sea. There are many variants, studies of which have estimated extremely low releases of activity to the biosphere and consequent hazard to man (see also the discussion to follow in Section 6.2.3). There are inherent advantages in disposal in saline groundwater. Safety factors appear so large that there is scope for relaxing the standard of some of the migration barriers. For example, with low-leach HEW in saline groundwater under the sea-bed, the nuclide pickup properties of the ground are not critical: disposal near the source of the waste, as by USL at Dounreay and Sellafield, could then avoid the extra cost and political opposition involved in transporting it elsewhere.

HEW and spent fuel are relatively concentrated heat sources, so that surface storage for several decades or dilution of the heat output may be considered to keep rock temperatures and/or excavation costs down. HEW is a low-leach waste form, but CMS, PCM and the more leachable spent fuels may require extra safety to compensate for their higher leachability, such as thicker packaging, low-permeability backfill or deeper emplacement.

Overall, there are a range of concepts for nuclear waste disposal with extraordinarily high safety factors, as discussed in Appendix 4.4 and Section 6.2.3; for application in the more distant future, elegant schemes can be conceived with groups of facilities on sandbanks, where nuclear plants decommissioned to Stage 3 could be left empty or used as Dry Boxes for LLW. Other wastes could be buried deep under the sandbanks; no sophisticated barrier arrangements would seem necessary.

It is important to distinguish between the treatment and disposal problems of the various nuclear fuel cycles of Chapter Two.

(a) The Thermal Recycle and Fast Reactor Cycles using reprocessed Magnox fuels create problems mainly from the resulting HAL and PCM and the Magnox cladding residues, either in silos or embedded in cement. These residues will occur only for a few more decades, until the phasing-out of Magnox reactors.

(b) Fuel recycling of AGR, PWR and Fast Reactor fuels does not involve the special difficulties with Magnox residues. The merits of future options of nuclear fuel cycles should therefore omit consideration of these residues.

(c) The main issue of Once-Through Cycle concerns the (as yet undemonstrated) disposal of spent fuel. Problems of reprocessing, in particular the disposal of most of the existing wastes at Sellafield, should not therefore be considered in assessing the merits of the Once-Through Cycle.

The solid wastes from coal power are far too voluminous for any alternative to simple dumping local to mines or power stations. The toxicants in the leachates from these and the liquid wastes themselves, transferred via local water flows eventually to the sea, present a long-lasting hazard greater and much more immediate than existing or projected nuclear waste disposal schemes. There are no calls for the containment of radioactivity from coal power dumps beyond a series of Ice Ages (as for nuclear wastes), in spite of the fact that the overall content of long-lived alpha activity in such dumps is at least ten times that of LLW (see Section 3.5.3). From this comparison, it is clear that criteria applied to disposal of wastes from nuclear and coal power are wildly inconsistent and hence it would seem reasonable to conclude that there is scope to relax the safety features in nuclear waste disposal to allow simpler operations such as disposal direct from the waste sources and shallow land burial of LLW above the water table at the coast.

Operational wastes from fusion power present less severe disposal problems than those from fission power because of the absence of fission products. The decommissioning wastes could well require similar processes to those from fission reactors. Renewable sources of power have few operational wastes and those created during construction and decommissioning should present little disposal difficulties. An exception is the tidal barrage, which, like the Stage 3 decommissioning residue of nuclear reactors, may never be deemed worthy of transferring to another disposal position.

Appendix 4.1

HEAT OUTPUT OF HIGH LEVEL WASTE AND SPENT FUEL

An important feature of burying heat-emitting radioactive wastes is that most rocks have only poor thermal conductivity. For this reason, the rock near to the waste can reach a temperature such that undesirable effects such as chemical decomposition or stress cracking occur. (The situation bears some similarity thermally to hay which has been stacked while damp, when bacterial action creates a relatively small output of heat; however the thermal conductivity of the hay is so low that eventually the stack reaches such a high temperature internally that it may catch fire.) Generally, a maximum temperature rise in rock of 100 °C is regarded as acceptable, though there are good reasons for believing that much higher temperatures would be safe, according to the type of rock.

We shall discuss the flow of heat through rocks in Appendix 4.2. Meanwhile, in preparation for this, we set out here the thermal characteristics of HEW. (For the first few hundred years ex-reactor, the thermal characteristics of spent fuel vary similarly with time.) Fig. 4.1 shows the variation with time of the heat output from such waste. The time is measured from the point at which the nuclides in the waste are no longer subjected to a reactor neutron flux, i.e. only decay processes are taking place: this time is commonly referred to as time 'ex-reactor'.

At 10 years ex-reactor, HEW has effectively lost nuclides of half-lives less than a few years: the next most important thermally are strontium-90 and caesium-137, half-lives about 30 years, and their short-lived daughters. Clearly, from the graph these can be seen to dominate the heat output up to about 150 years ex-reactor, when the isotopes of americium, mass 241 and 243, and of plutonium, mass 239 to 242, take over. Arrangements to provide cooling for the waste for (say) 150 years are readily feasible by modern engineering. A useful drop in heat output by a factor of about ten is thereby obtained, which, in turn, reduces the volume of rock that needs to be excavated in order to spread the thermal load through the rock satisfactorily. On the other hand, the half-lives of the americium and plutonium isotopes are so long that surface cooling to decay out an appreciable fraction of these is not really practicable.

Appendix 4.2

HEAT TRANSFER CONSIDERATIONS IN NUCLEAR WASTE BURIAL

The flow of heat through solids is governed by the rate of temperature drop with distance through the solid, the area of heat flow and a constant defining the inherent ability of the solid to conduct heat, defined as the thermal conductivity. If we consider the section of solid shown in Fig. 4.13(a), area A square metres, temperature drop T_2-T_1 °C across a thickness x metres and a thermal conductivity k in units of calories/sec/°C/metre, this can be written

Rate of heat flow across the solid =
$kA(T_2-T_1)/x$ calories/sec.

Most waste emplacement positions will be boreholes or tunnels, through the walls of which most of the heat will flow. An end view of such a circular cavity is shown in Fig. 4.13(b). Clearly, the area across which the heat flows increases as the distance from the wall of the cavity increases. This is complicated by the fact that the output decreases with time as the nuclides decay. However, all such characteristics can be set down as mathematical equations and solved. Using our constraint of Appendix 4.1 that the maximum temperature rise in the rock (which will occur at the cavity wall) is 100 °C, sets of answers can be obtained for a variety of assumptions of conductivities, cavity radii and times of disposal.

General inferences, perhaps rather obvious, are:

(a) the smaller the cavity radius, the lower the heat load per unit length that can be placed therein, and

(b) the more rapid the decay of heat output, the sooner the maximum temperature is reached.

In the particular conditions of the UKAEA study (Burton, 1981), it was shown that the shielded containers of vitrified waste (overall length and diameter 3.66 and 1.13 metres respectively) could be packed end to end in a close-fitting borehole at about 70 years ex-reactor or in a 3-metre radius tunnel at about 40 years ex-reactor (see Fig. 4.13(a) and (b)). If it were thought desirable, 'dilution' of the heat output could be achieved by interposing material such as steel between containers, which would spread out the heat over a greater length (Fig. 4.13(c)). Such 'dilution' would allow early burial but would obviously increase excavation requirements and so its extra cost would have to be weighed against benefits of decreased surface storage.

Figure 4.13: HEAT TRANSFER IN RADIOACTIVE WASTE BURIAL

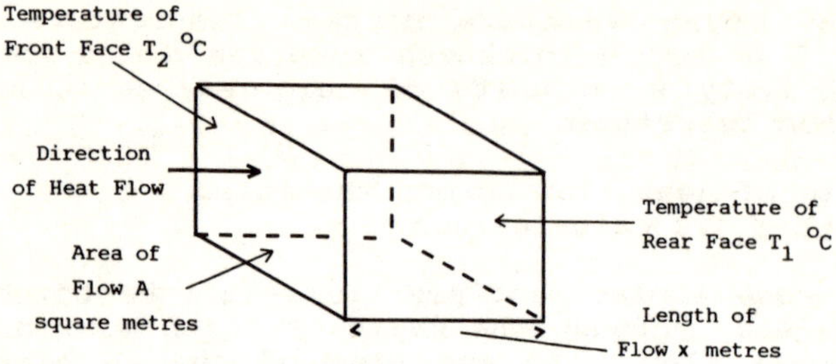

(a) PRINCIPLES OF HEAT FLOW

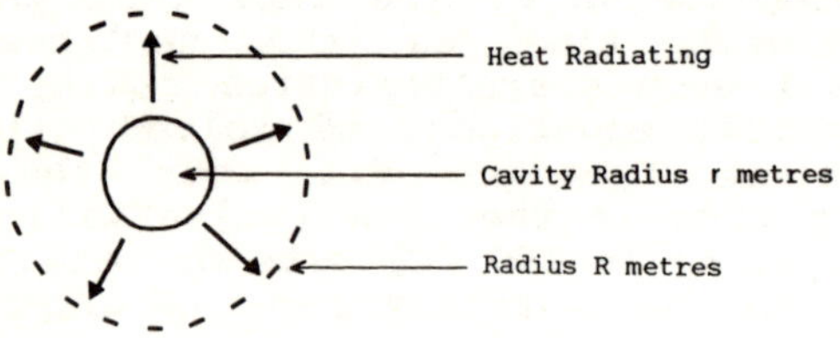

(b) RADIAL FLOW OF HEAT IN END VIEW OF BOREHOLE OR TUNNEL

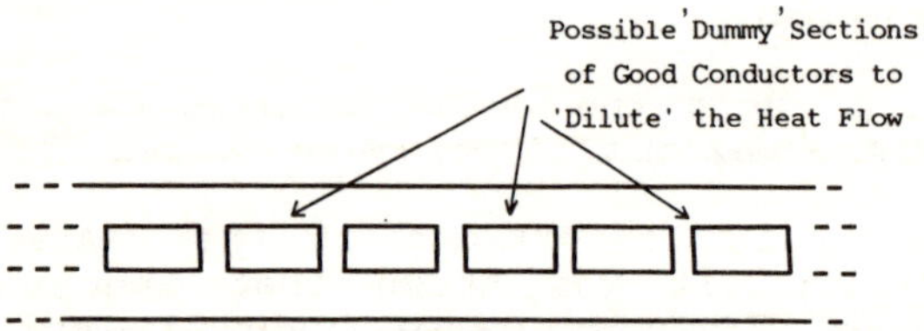

(c) PACKAGING END-TO-END IN TUNNEL

(d) SPACING OF BOREHOLES TO PREVENT THERMAL INTERACTION

Small-diameter boreholes, say at 0.2 metres, are of interest because it is feasible to drill them to great depths. Current sizes of HEW packages would be too bulky, but liquid waste and spent fuel could, in principle, be transferred down such narrow holes (see Section 4.1.4). The extra safety afforded by great depth might be regarded as an acceptable alternative to more sophisticated barrier systems.

Finally, thermal interactions between waste packages must be estimated. If the spacing is too close, the thermal effects from several positions can operate on some points in the rock, the overall temperature rise being the sum of the individual effects (Fig. 4.13(d)). The UKAEA study (Burton, 1981) found that, for the condition of packages assumed, emplacement in boreholes needed to be at least 20 metres apart and tunnels at least 60 metres apart, in order to avoid such additive effects. An interesting consequence of this was that, in order to achieve such separation for a borehole system, the connecting tunnels needed to be as long as the simple tunnel-only system. In this case, the boreholes became an additional excavation requirement. In the case of spent fuel disposal, the long-lived plutonium heat output could make adequate separation of emplacement positions even more important.

Appendix 4.3

HYDROLOGICAL ASPECTS OF NUCLEAR WASTE DISPOSAL

Fig. 4.10 shows highly simplified versions of natural water flows, (a) inland, and (b) at the coast. In both cases, part of the rain falling on higher ground runs off to streams on the surface and another part evaporates after reaching the ground. The rest percolates through the ground more or less vertically until it meets the water table, below which the voids within the rock are filled with water. The direction of flow is then sidewards and downwards as indicated by the flowlines.

In the inland case, the water may eventually emerge from the ground miles from its initial percolation point, either at the ground surface or under surface water. Near the coast, there is a changeover to saline groundwater under the sea-bed, with an interface between the two types of water. The reason for this is that the saline water is denser than the freshwater, exerting a higher pressure for the same water depth. Accordingly, seawater tends to penetrate under the land until the head of the inland water table has counteracted the density effect. Since the freshwater cannot pass the interface, it has to emerge at the surface near the top of the interface.

The practical situation is always more complex than described above. Less permeable strata can force springs to appear on nearby hillsides or direct freshwater to emerge hundreds of metres out to sea - the interface itself is never sharp. However, the simple picture enables four types of disposal location, defined in Section 4.1.8.1., to be outlined as in Fig. 4.10. Type (a), above the water table, is depicted as point A in Fig. 4.10(a). Types (b), (c) and (d) are depicted by points B in Fig. 4.10(a), with C and D shown in Fig. 4.10(b). At point A, waste would only possibly be contacted by percolating water; radioactive waste under the water table at point B in (a) would be in a region of flowing fresh groundwater, whereas, under the sea-bed seaward of the interface at point C in (b), because of the flat sea surface above, there would be no difference of water head to make the groundwater move. Moreover, any long-term diffusion of nuclides from C must emerge in and be vastly diluted by the (undrinkable!) sea. There would appear to be less merit in disposal at D in Fig. 4.10(b) when compared with C, because groundwater is flowing.

Examples of possible USL disposal of particular importance occur at Dounreay and Sellafield. A few tens of metres from the Dounreay shore is a crushed zone, encountered during the construction of the effluent tunnel (Shimmin, 1963). This could ensure easy flow of freshwater into the sea, were it even to permeate this far through sea-bed strata. A repository could then be constructed seaward of the crushed zone in saline groundwater. For practical purposes, access tunnels might cross the zone at fairly shallow levels for ease of construction, then be directed to acceptable disposal depths.

At Sellafield, the emergence of freshwater well out under the sea-bed is possible, though from existing geological data, less than likely. Exploration to establish satisfactory disposal locations could be expensive. On the other hand, the associated costs are unlikely to be high relative to (say) the expense of packaging the bulk of UK nuclear waste (which arises at Sellafield) and transferring it for disposal at Dounreay.

Appendix 4.4

THE MIGRATION OF RADIONUCLIDES IN GROUNDWATERS

The movement of groundwater following percolation of rainwater and the development of varying water table heights has been outlined in Appendix 4.3. Another source of water movement is caused by convection after the lowering of water density round a repository from which heat is emitted. The surrounding water then becomes buoyant and tends to rise and circulate, after the manner of many unpumped household central heating systems (Fig. 4.12(a) and (b).

Any radionuclides in such moving groundwaters will tend to move in the same direction as the water. However, there will be mechanisms reducing their mobility and concentration. Firstly, minerals in the rocks, through which the water passes, pick up radionuclides by physical and/or chemical reactions. Though the effect is normally reversible, the overall result can be a considerable hindrance to the migration; in calculations of migration phenomena, a parameter K, known as a retardation coefficient, is often defined, which is effectively the factor by which the nuclide movement is slower than the surrounding groundwater. In many cases, depending on the nuclide, the chemistry of the groundwater and the minerals in the rocks, K can have a value of several thousand, e.g. K for plutonium moving through clay has been estimated as 3,500 (Hill, 1978). The second effect is that of dispersion of the nuclide, i.e. a degree of mixing caused by the tortuous route of the water through the cracks and pores of the rock. This tends to smear out the concentration of the nuclide, so that the escaping 'front' is not sharp. Estimates may therefore be made of the probable migration behaviour of nuclides. Practical confirmation is, of course, difficult to demonstrate, but the natural reactor at Oklo (Section A.1.1.2) has shown how little actinides and fission products have moved in hundreds of thousands of years.

In addition to the migration caused by moving groundwater, radionuclides can move in stagnant water, e.g. under the sea-bed, by diffusion. Here, the water molecules in the groundwater collide with ions or molecules containing the radionuclide, inducing a random movement, so that eventually the radioactivity spreads in all directions, obviously becoming more dilute with distance from the source. Superimposed on this diffusion is the retardation effect described above.

All of the above effects can be represented in mathematical equations. Simplifying assumptions, which are 'pessimistic', i.e. tend to exaggerate the rate of migration, can be made so that solutions to the equations can be found reasonably quickly and easily (Fig. 4.12(c)). For example, if migration is assumed to follow only one direction with no losses or diversions, answers can often be derived readily using microcomputer programs. These answers are a very useful guide in early comparisons of outline concepts; much more sophisticated calculations may be deemed worthwhile as a concept is firmed up. It must be emphasised, however, that every cubic metre of rock cannot be investigated to provide data for the calculations and so the results are no more reliable than the assumptions. It is important, therefore, that where these are somewhat uncertain, they must be chosen to be pessimistic; in practice, several levels of parameters are often put into the computer runs to see how important the uncertainty in the parameters is to the conclusions reached from the calculations.

Many thousands of computer hours have been spent in estimating the efficiency of retention of radionuclides within given repository conditions. Most of these results suggest very high degrees of safety indeed (see, for example, studies of disposal in flowing groundwaters in Burton (1981) and Hill (1978)). If activity were to escape to the land surface before entering the sea, rather arbitrary estimates have to be made about human conditions in the distant future. The National Radiological Protection Board (NRPB), for example, assumes that the release of activity would be into drinking water flowing at a rate of 0.3 metres per second, supplying a town of 30,000 people (Hill, 1981).

On the other hand, the facts that diffusion processes are so much slower than those of convection, and that seawater cannot be ingested, imply there are considerable advantages in disposal under the sea-bed. Here in the UK, there are few locations far from the sea and the most important nuclear waste sources are also near the sea. It would appear prudent, therefore, to take advantage of this favourable geography and not closely follow disposal concepts proposed in the US, where the distances from the waste source to the coasts are considerable.

A result of interest from NTC computer runs in this respect is that a negligible fraction of plutonium can migrate by diffusion through 100 metres of clay, whereas 500 metres of granite would be needed to provide the same effect. Neptunium, which in general is less strongly picked up by minerals, would need several times the above thicknesses for full retention. Nevertheless, its release at all times would be extremely slow, so that, if emerging into the sea, there would be no detectable local concentration at any time. (Appendix 6.2 extends this assessment to show that there would be no significant hazard at any time.) Calculations quoted in Bury (1985) suggest broadly the same picture. These evaluations suggest that HEW and spent fuel could be buried safely beneath the sea-bed, without any sophisticated pretreatment such as vitrification. Sufficient safety would be better obtained simply through extra depth rather than expensive and complex additional barriers. It is then a matter of judgement whether to bury such waste close to its source as under the sea-bed at Sellafield, or (say) to transport it to a possibly more favourable geological position in the clay beneath the sandbanks of the Wash.

Certainly, there would appear to be little merit in disposing waste under the bed of the ocean compared with disposal in saline groundwater close to the coast. Both systems will allow only extremely slow if not negligible release of nuclides to sea. No local effect will therefore occur in either case and the overall effect globally will be negligible. There would therefore seem to be no case for choosing the more difficult and expensive procedure. It is worth bearing in mind, too, that the top 500 metres of land in Britain contains 300 tonnes of radium and 800 million tonnes of uranium (Appendix 1.1)! Most of this can migrate more readily to the surface than nuclear waste placed in a carefully selected location at (say) 500 metres depth.

Appendix 4.5

POTENTIAL METHODS FOR THE DISPOSAL OF HAL BENEATH THE SEA-BED

The manner in which the heat output of spent nuclear fuel falls with time is dominated by the fission products for a century or so (see Fig. 4.1 and the discussion in Appendix 4.1). If spent fuel is buried in rock, say a borehole, the adjacent rock wall temperature will rise at first, then fall as the heat output falls and the heat dissipates further through the rock. The wall temperature therefore passes through a maximum, which is clearly unaffected by the physical state of the heat source, merely the initial output and the form of its decay with time. Since HAL and HEW contain the bulk of the fission products, we can use the same heat decay curves as spent fuel to estimate maximum rock temperature increases, whether the disposed form of waste is HAL, HEW or spent fuel. The particular case of HAL disposal is now considered: deep burial of spent fuel and HEW will be described further in Appendix 4.6.

Using the decay curve in Fig. 4.1, it may be estimated (NTC, 1986a, and Appendix 4.2) that if 50 cubic metres (corresponding to 2.5 GW(E)yrs of power) at 4 kilowatts per cubic metre of typical Sellafield HAL were placed in the bottom 500 metres of a borehole, with diameter 0.34 metres and 1,000 metres long, then the maximum rock temperature rise would be 100 °C, a level generally regarded as acceptable. Alternatively, if the waste were diluted by a factor of two with preferred additives, the rise would be 50 °C. In addition, if the diameter of the hole were increased to 0.50 metres, the rise would be 100 °C. The use of smaller boreholes with the same waste characteristics would reduce the rock temperature rise but increase the number of boreholes required, since halving the diameter means four times as many holes are needed to obtain the same volume.

Since HAL is denser than seawater, the system would be physically stable (though possible secondary effects from gases formed by radiolysis of the liquor would need careful assessment). The introduction of liquor would, of course, be at the bottom of the borehole to displace the seawater upwards before the hole was plugged above the liquor. After emplacement of the liquor, two scenarios can be imagined. Firstly, if the rock around the borehole contains fissures, the liquor will spread outwards and downwards (because of its higher density than the surrounding groundwater in the fissures (see Fig. 4.2(i)). The liquor cannot cross the saline/freshwater interface (again because of its higher density). Probably by reaction of the acid in the liquors with minerals such as calcite, eventually most of the activity will precipitate out and the fissures become blocked. Suitable chemicals could be added either in the liquor or as a preinjection to enhance this effect. This type of scenario could well occur in the Kirklinton Sandstone which, from a borehole log at Seascale (Gregory, 1915), probably extends roughly one kilometre below the sea-bed at Sellafield. Because of the stagnancy of the groundwater, little horizontal movement of the disposed liquor will occur. The second possibility is that the liquor remains substantially within the borehole. This could well happen naturally in the Borrowdale Volcanics which exist below the Kirklinton Sandstone at Sellafield (see Fig. 4.2(ii)), though occasional fissures are possible even here.

If some extra stabilisation were required, the borehole could first be filled with a chemical solution such as a mixture of sodium hydroxide and silicate, followed by dropping active liquor through the first solution, so that three-dimensional solid networks would be formed, enclosing the activity and blocking the fissures. Such concepts have not been tried at great depths, though a form of such solidification is known in shallow burial waste treatment as 'Chemfix' (Env. Sci., 1973). Early tests could be done on medium-active liquid wastes, say containing ruthenium-106, which would then no longer require expensive holdup tanks or discharge to sea (Section 3.2.4). The only locations suitable for the disposal of such liquors because of the difficulties in transporting the latter are the sites where they originate, i.e. Sellafield and Dounreay. The latter site could be particularly advantageous, as a crushed rock zone just off the coast (Shimmin, 1963) might behave as a natural 'borehole' of considerable diameter and depth. At Oak Ridge in Tennessee, US, liquid radioactive wastes have been dispersed for many years by mixing with cement grouts and forcing open shale strata wherein the grout sets. However, the water pressure used to open the strata can cause uplift at the ground surface, perhaps affecting nuclear plant. The obvious alternative of mixing liquors with cement before passing into a borehole could run into difficulties if premature setting of the cement in the pipeline occurred. On the other hand, injection of a viscous gel is a well-known operation (see, for example, Batchelor (1985)); radioactive waste might be passed in such a gel to the bottom of a borehole, followed by a 'pipecleaning' inactive gel above it, finally cementing in the top section.

Appendix 4.6

THE FEASIBILITY AND COSTS OF DEEP BOREHOLE DISPOSAL OF NUCLEAR WASTES

A.4.6.1 General

Most of the deep land burial studies to date have been based on reaching the disposal zone for the nuclear wastes by access tunnels. Emplacement of waste packages is then achieved by either spacing them inside disposal tunnels or lowering them down boreholes in the tunnel floors. This Appendix examines the general feasibility of constructing deep boreholes from the ground surface with packages emplaced in the lower sections and subsequent sealing of the upper levels. An indication of the costs of borehole preparation is given in each case to check that these are not so high as to render a given scheme uneconomic.

A.4.6.2 The Geometry of Spent Fuel and Nuclear Wastes with Respect to Borehole Emplacement

A PWR providing 1 GW(E)yr or 8.8 billion kwhrs of electricity will discharge on average 50 subassemblies containing 30 tonnes of spent fuel. If the spent fuel is reprocessed (thereby substantially increasing waste volumes), there would be:

(a) 4,000 drums of LLW, each of volume 0.5 cubic metres, (say) diameter 0.8 metres and length 1 metre,

(b) 220 drums of ILW, each of volume 0.5 cubic metres, (say) diameter 0.8 metres and length 1 metre, and

(c) 20 cubic metres of HAL, or, alternatively, if this is vitrified,

(d) 15 AVM containers each containing 0.18 cubic metres of glass at 15% fission products by weight, diameter 0.42 metres and length 1.3 metres.

A.4.6.3 Boreholes of Diameter 0.3 Metres

Armstead (1983, p.294) gives over twenty examples of boreholes with diameters around 0.3 metres in a variety of rocks at depths ranging from 0.5 to 8 kilometres. From this set and data from UK drilling experience (Allott, 1986), the most expensive type, in hard rock, would cost in the region of £600,000 for a depth of one kilometre. A diameter of 0.3 metres could accommodate a PWR sub-assembly, which has a square cross-section of side 0.21 metres, plus a probable overpack. Allowing a length of 5 metres of borehole per sub-assembly and sealant, say of cement, the maximum rock temperature rise would only be about 30 °C, even if the fuel were disposed at 10 years ex-reactor. 100 sub-assemblies could be accepted in the bottom half of a 1-kilometre hole. This corresponds to a borehole cost of £6,000 per sub-assembly or the equivalent of 0.003p per kwhr. Extrapolations of costs for hard rock, also given in Armstead, suggest that if the bottom 2 kilometres of a 4-kilometre hole of the same diameter were used for sub-assembly disposal, equivalent costs would rise to about 0.01p per kwhr.

A 500-metre length of a borehole of diameter 0.3 metres would have a volume of about 35 cubic metres, equivalent in terms of HAL to about 1.8 GW(E)yr or 15,000 million kwhrs of electricity. Using the bottom half of a 1-kilometre borehole for HAL disposal would then have an equivalent cost of about 0.004p per kwhr.

A.4.6.4 Boreholes of Diameter 0.5 metres

If more packing or barriers were thought desirable round the above PWR sub-assemblies, a larger diameter, say 0.5 metres might be preferred. From Armstead and sources in the drilling industry, it seems reasonable to assume that borehole costs are roughly proportional to their diameter: this increases the costs for disposal in a 1-kilometre hole from 0.003 to 0.005 kwhr.

Since the volume per unit length of hole depends on the square of the diameter, it is readily seen that the bottom 500 metres of a 0.5-metre diameter hole would have a volume of about 100 cubic metres, giving a borehole cost for HAL disposal of about 0.002p per kwhr. This would give too high a rock temperature rise, so either the HAL (at 10 years ex-reactor) would have to be diluted, increasing borehole costs proportionately, or the HAL would have to be stored for a few decades before disposal.

Allowing 1.5 metres for an AVM container plus sealant, some 330 containers could be accommodated in 500 metres of a 0.5-diameter hole. At a cost of £1M for the 1-kilometre hole, we arrive at £3,000 per container or 0.0006p per kwhr. Containers would have to be stored for about 50 years to reduce rock temperature rises to an acceptable level.

A.4.6.5 Boreholes of Diameter 1 Metre

Costs for the above larger diameter are not readily available in the literature, but using the previous linear dependence on diameter it would seem reasonable to take £2M as the borehole cost in hard rock. About 8 PWR sub-assemblies could be packed into a container which could be lowered down a 1-metre diameter hole, i.e. about 800 in the bottom half of our basic 1-kilometre hole. This corresponds to a cost per sub-assembly of £2,500 or 0.001p per kwhr. Storage of several decades would be necessary to keep rock temperature rises acceptable.

It is readily deduced from the previous examples that HAL costs in the standard hole of 1-metre diameter would be 0.001p per kwhr. If the hole were a little larger, perhaps 3 containers of AVM glass could be fitted side by side; the cost in this standard case would then be about 0.0004p per kwhr.

With a small allowance for sealant between drums, perhaps 400 LLW or ILW packages could be accommodated in the standard hole at 1-metre diameter. The associated borehole costs are then £5,000 per package (or 0.23p per kwhr for LLW and 0.01p per kwhr for ILW).

A.4.6.6 **Summary**

These very rough estimates indicate that deep borehole preparation costs are unlikely to add significantly to electricity costs in the cases of disposal of spent fuel, HAL, HEW (AVM glass) and ILW. In some cases, reduction of the heat output by dilution, or storage for several decades, might be preferred to decrease heat rises in the rock or the use of larger-diameter boreholes. On the other hand, such disposal for LLW does not look attractive.

It must be reiterated that the above costs only refer to the non-radioactive operations of borehole construction and sealing: costs of equipment and operations to carry out very deep lowering safely have not been evaluated.

Chapter Five

HEALTH HAZARDS FROM POWER SYSTEMS

5.1 GENERAL

Casualties from the operation of power systems are of two main types: physical injury and the effects of various diseases. The first of these requires little explanation and is therefore recounted with various accident statistics in Chapters Six and Seven. The present chapter outlines the manner in which the wastes described in Chapters Three and Four can interact with the human body, firstly through radiation and secondly through chemical effects. The nature of the hazards due to these interactions are then discussed in a third section, followed by a summary of the chapter in the fourth and last section.

5.2 THE INTERACTION OF RADIATION WITH HUMAN CELLS

In Section 1.2.1, the various types of radiation associated with nuclear power were introduced. Of these, neutron radiation effectively occurs only in a reactor: the various decay processes of radionuclides discharged from a nuclear site either routinely or by accident contain few neutron emissions. It is perhaps also worth emphasising that, for all practical purposes, decay radiation does not induce radioactivity in other nuclei: the decay merely involves a particular nucleus undergoing changes before it achieves stability, the associated radiation interacting either occasionally by elastic collisions with other nuclei or more often displacing electrons outside the nuclei, thus changing chemical characteristics. An important consequence is that the packaging round nuclear waste or backfill during disposal does not itself become radioactive. On the other hand, the physical transfer of activity, (say) by leaching of radionuclides from waste into backfill or the pickup of reprocessing chemicals on equipment, can occur and is known as contamination. Radioactivity is also brought into contact with man by coal-fired power, not by its creation by fission, but by redistribution of natural radionuclides excavated with coal and either emitted with stack gases or dumped with fly ash or spoil.

The helium nuclei comprising alpha radiation lose energy relatively easily when encountering human cells. Consequently, atoms can be displaced and electrons removed from the latter in a small region, i.e. the damage is highly localised. On the other hand, the fast-moving electrons comprising beta radiation lose little energy on collision with the nuclei of atoms, since their masses are relatively so very small. Displacement of atomic nuclei in the human body is therefore insignificant. Consequently, energy is lost more gradually than for alpha particles, mainly by reaction with electrons outside nuclei. Beta radiation therefore leaves a more diffuse string of chemically altered cells behind it.

Gamma radiation interacts even less readily with human cells than beta radiation, but when its energy is transferred to the electron, the latter behaves like beta radiation. The overall effect is therefore like beta radiation but spread over a larger volume.

In considering the effects of radioactivity from power systems, it should be noted that these take place against a background of radioactivity in the environment. Man is continuously bombarded by natural gammas. For example, uranium and thorium decay systems give off associated gammas in soil and buildings, gammas given off by potassium-40 in blood and tissue irradiate the body internally; cosmic rays from space contain a component of high-energy electromagnetic radiation, much of which can penetrate through buildings and the human body, and X-rays, which are effectively low-energy gammas, are directed at various parts of the body in medical tests. Natural radiation strikes each cell of the human body many times every year (Fremlin, 1986). There is, however, a rapid and effective mechanism to repair the corresponding damage and this obviously cannot distinguish between whether the changes were caused by natural or power system radiation.

Damage to the human body from radioactivity can occur in three basic ways. Firstly, direct radiation from outside the body may penetrate cells. As explained above, the range of alpha radiation in solids is very short; even a sheet of paper is sufficient to stop the helium nuclei. Since any solids discharged from nuclear sites will be surrounded by material more effective as a barrier than paper, direct radiation hazards from alpha radiation are unimportant. For a similar reason, it is not very likely that external irradiation from betas will be dangerous and so the predominance of alpha and beta damage in the body arises through contamination internally. Gamma radiation is often emitted with beta decay: an important example of this is the decay of cobalt-60 to nickel-60, where the beta particle is accompanied by penetrating gammas. Such material must therefore be surrounded by adequate shielding. Except for low-probability events, as in the re-excavation and removal of shielding from a waste package in the distant future, or to operators in a severe accident like Chernobyl, direct radiation from nuclear power systems is of little concern. Similarly, radiation emitted from coal power dumps can be considered innocuous.

The second type of radioactive damage to man is through ingestion, i.e. via contaminated food and/ or drink. If the retention time of radioactivity within the body is short, there are no significant ill-effects. However, if the radionuclides are assimilated into the body, they may remain there for years. Localised damage to vital organs can then be serious, particularly for alpha emitters, whose concentrated disposition of energy at some locations can overwhelm body repair mechanisms there (Fremlin, 1985). The chance and consequences of ingestion of radionuclides are important features in comparing the safety of nuclear and coal power waste after disposal.

The third mode by which man may be affected by radiation is through inhalation. Some radionuclides pass in and out of the lungs with little effect. An example of this is xenon-133, half-life just over 5 days, which is discharged from reactor stacks. Being an inert gas, it is very unreactive chemically and so is not retained in the body; moreover, its decay product, caesium-133, is a stable nuclide and virtually innocuous.

Another radioactive gas which is of particular importance to the public because of its accumulation within buildings and discharge from coal power operations is radon. Two isotopes of this occur naturally from the decay of uranium and thorium in soils. Like xenon-133, radon is an inert gas and so is unreactive chemically, so that, if inhaled, it is soon expelled. However, the high-energy daughter products (see the decay chain diagram in Fig. 1.1) can attach themselves to particulates in the air which are not exhaled easily. The basic cause of uranium miners' lung diseases is adsorption of polonium isotopes formed from radon on to mine dusts (Patterson, 1983): concentrated radiation energy is then deposited within the lung (RCEP, 1984).

5.3 CHEMICAL EFFECTS ON HUMAN CELLS

5.3.1 General

The effects of chemical pollution from power systems on the environment are extremely complex and in many respects not well understood. The main two routes for chemically toxic effects on man, just as from radiation, are ingestion and inhalation. Individual chemicals often seem to cause little damage on their own, but in the presence of other materials, deleterious effects can occur. Even more indirect effects can be produced, in that pollutants can alter the chemistry of the environment, such that normally innocuous natural materials become toxic. The discharges of chemicals from nuclear and renewable systems are slight (Tables 3.1 and 3.3) and so their effects are insignificant: in the section below, toxicants in the substantial discharges from coal-fired power systems are divided into three classes - organics, trace metallic elements and large-scale inorganic wastes.

5.3.2 Organics

The most important source of organic toxicants arises from the incomplete combustion of coal. Since the latter itself contains many forms of hydrocarbons, part of these can be discharged unchanged or only partly oxidised. The mixture is so complex that it is often known under the acronym BAP, derived from an important constituent, benzo-alpha-pyrene. This is a carcinogen also present in tobacco smoke: much research has therefore been done in establishing its toxicity and reasonable estimates of casualties from its inhalation can be made (Comm. En. Env., 1981).

5.3.3 **Trace Inorganics**

The elements arsenic, beryllium, cadmium, chromium, lead and nickel are all present in small concentrations in coal and their carcinogenic properties have been well established, in particular by the US Environmental Protection Agency. Cohen (1983) has used these studies to estimate the risk per gramme of these elements ingested via food. Arsenic, for example, is respectively about 5 and over 10,000 times more toxic per gramme ingested than plutonium and neptunium (see also ICRP, 1986). From a knowledge of their concentrations in coal and their emission rate from power station stacks, Cohen was able to deduce the amount taken up in food of these trace inorganics before they are leached out of the ground and transported, via streams and rivers, out to sea. Once there, their situation corresponds to that of radionuclides migrating from a disposal position where groundwater eventually transfers them into the sea. A rough comparison of the toxic potential of the two types of waste can then be done, but this can only be generally indicative of relative hazards, since the pathway to human consumption, usually via fish, is uncertain for most non-radioactive toxicants. Indeed, far more is known about the consequences of radionuclide discharges than for many effluents containing toxic non-radioactive chemicals.

Spoil heaps, from coal excavations, and lagoons, from coal cleaning and fly ash settling, contain toxic chemicals such as compounds of mercury, selenium and vanadium, albeit at low concentrations. Leaching removes about 2% of the immense volume of the heaps (see the annual arisings in Table 3.2). This and the overflow from lagoons can therefore transfer about a million tonnes per year of dissolved chemicals (including the radionuclides uranium and thorium and their decay products) at the current level of coal-fired power in the UK, through surface and groundwaters eventually to sea. Ingestion of food contaminated by leachates results in the accumulation of these chemicals in body cells and causes associated damage.

5.3.4 **Large-Scale Inorganic Wastes**

Discharges to atmosphere from coal-fired power stations contain huge quantities of particulates. A 2 GW(E) power station discharges about 20 tonnes per day of particulates below 0.02 millimetres in diameter (Comm. En. Env., 1981). Unfortunately, some of the particles, which are small enough to pass through the filters, are thereby likely to penetrate through the body defences into the deep regions of the lung, where they are trapped. The percentage by weight taken out by the filter does not therefore give a correspondingly large reduction in hazard to the lungs. The US Environmental Protection Agency (1977, 1979, 1980 and 1981) suggests that particulates may not be specially hazardous unless accompanied by other chemicals which are absorbed on them. Chappie (1982) and Ozkaynak (1985), for example, establish that sulphate and fine particle levels are closely related to mortality from air pollution. Possibly sulphur dioxide absorbed on a particle gives a very high local concentration or 'hotspot' when in contact with respiratory cells, overwhelming body defences at that point. Oxides of nitrogen, too, are known to cause health problems. However, it is difficult to separate out the main causes and mechanisms of detriments from stack discharges because of

(a) the complexity of the pollutant mixture and interactions between its components,

(b) the long timescale over which ill-effects occur,

(c) the general loss of health which may enhance the likelihood of death from other causes, and are therefore attributed wholly to the latter, and

(d) the contribution from other sources of pollution, mainly from road vehicle exhausts.

A secondary and yet very important effect of the immense discharges of oxides of nitrogen and sulphur (Table 3.2) occurs because stack heights have been increased considerably in the last few decades to reduce the severity of local fall-out. The oxides can then be transported considerable distances by wind before returning to ground, often as rain. The nitrogen and sulphur are then in the form of acids: in addition to massive damage to buildings (Barret, 1973; Day, 1985), these acids can change the chemistry of ground and surface waters. The overall effect has become known as 'Acid Rain' or 'Acid Deposition' (RCEP, 1984); some of the biological consequences are to exterminate birds and fish through increased leaching of aluminium into streams, rivers and lakes, following enhanced acidity in natural waters (Price, 1983), to decimate whole forest areas, and to exacerbate aluminium dissolution into drinking water, possibly resulting in brain damage and deaths from senile dementia (ITV, 1986). A recent review by Pearce (1987) gives an excellent description of the appalling consequences of burning fossil fuels, much of which is from coal-fired power stations. The situation is succinctly set out in the introduction to his book as below.

> The skies above Europe are poisoned. Toxins are carried on the breezes from power stations and autobahns across the most polluted continent on Earth. When the poison falls to the ground, it chokes the pores of leaves on trees from the Alps to the Urals, it eats away at stone and brick, paper and rubber; it destroys soils and flows into rivers where it kills fish by disrupting the operation of their gills. It kills humans, too ... Acid rain is the phrase everybody uses to describe this poison. But rain is only part of the story. Acid mists and fogs are even more dangerous. London's most famous pea-souper, the smog of 1952, killed several thousand people. We now know that the water droplets in that fog were nearly as acid as the water in a car battery. Today, scientists are finding mists almost as acid on Scottish hillsides. Nowhere is safe ... The chimneys and exhaust pipes of Europe are creating an ever more complex cocktail of chemicals in the air over the Continent. Some react with sunlight to form ozone, a

> chemical which damages trees and crops and irritates the human lung. Ozone also speeds up the conversion of other gases to Acid Rain. Countries such as Britain and West Germany have banished smoke. But clearer skies only make the cocktail more reactive and increase the threat from ozone and acid rain. The familiar heat haze seen on any sunny summer's day is made up of acid particles created by ozone ... All of this should frighten us. Our forests and fish, cathedrals and crops, lichen and lungs - all are under attack. Today's air pollution is every bit as lethal as the black smogs it has replaced. And it is everywhere, from the westerly shores of Ireland to the lakes of Scandinavia.
>
> (Pearce, 1987)

The link between coal-fired power stations and the above effects of Acid Rain over much of Europe has now been accepted by the UK Government and equipment to reduce sulphur and nitrogen discharges will be installed at three large power stations (announced in 1987). However, this will be expensive and only provide a reduction of about 7% in total UK sulphur emissions. Moreover, in recent years, as recounted above by Pearce (1987), the increase in ozone levels has been suspected of causing serious health effects. In the long hot summer of 1976, several hundred people died above the average rate in London, corresponding to the worst ozone period on record in Britain. Overall, both the direct and indirect effects of huge quantites of inorganic chemicals discharged via coal-fired power stacks are very serious. Even Friends of the Earth writers describe the CEGB as 'the biggest single polluter in Europe' (Porritt, 1984).

Finally, a third effect from coal combustion, which may become the most serious hazard of all, arises from the enormous quantities of carbon dioxide discharged. Its concentration in air may consequently double in the next 50 years because of the increasing rate of burning fossil fuels: at the same time, the destruction of large numbers of trees, as in the rain forests of South America, could reduce the conversion of carbon dioxide back to carbonaceous materials and oxygen. In some ways, the increase may be beneficial. However, the main concern is over an enhanced 'Greenhouse' effect. This is so-called after the trapping of radiation in a greenhouse, where light passes through the transparent cover and is then partly re-emitted as infra-red radiation, after striking the ground. The cover is not transparent to this longer wavelength radiation, which therefore remains as heat to increase the temperature inside the greenhouse. Atmospheric gases such as carbon dioxide, ozone, methane, nitrous oxide and chlorofluorocarbons absorb infra-red radiation emitted after sunlight strikes the Earth's surface. Increased levels of these consequently increase the temperature of the air, the greatest individual effect coming from carbon dioxide (McElroy, 1988). The associated change in world climate could reduce crop yields; perhaps an even more serious consequence, however, is the melting of the West Antarctic Ice Sheets, a hazard accepted by FOE (Flood, 1987). This could raise the level of the oceans 5 to 8 metres, flooding many low-lying major cities (Greenhalgh, 1980). It has been commonly assumed that the above changes in climate and their consequences would be gradual. However, recently, there appears to be clear evidence that effects could be sudden and consequently much more serious (Broecker, 1987). Moreover, the effects may not be uniform, possibly increasing rainfall in the UK but decreasing it leaving dustbowls in the US Mid West: monsoons may fail in India, bringing widespread famine (BBC2, 1988). It is possible, too, that the Greenhouse effect and the 'hole' in the ozone layer, which has developed over the Antarctic in the last few years, are interconnected (McElroy, 1988).

5.4 RESULTANT HEALTH EFFECTS FROM POWER SYSTEM DISCHARGES

5.4.1 Risk and Consequence

In assessing the effects of power system emissions on the human body, it is useful first to divide the possibility of harm into two parts, risk and hazard. As used generally in the scientific literature and hereafter in this book, hazard is a set of circumstances which may cause harmful consequences and the chance of it doing so is the risk associated with it. A car travelling through a town is a hazard which may injure or kill children; the risk of this hazard increases with the number of children near the path of the car.

Most toxic chemicals, such as carbon monoxide, are generally believed to require a 'threshold' exposure before any harm is incurred. On the other hand, when the hazard (from either radiation or ordinary chemicals) is a form of cancer (and probably for some non-cancerous diseases, too), a linear, no-threshold relation is assumed between exposure and the risk of induction of disease. An important consequence is that a given intake of a carcinogen initiates the same number of cases whether it is spread over a small population or a large one.

In the sections to follow, the above assumptions will be followed, though there is no direct proof of the linear hypothesis and an increasing number of experts are sceptical of its validity. The numbers of cancers quoted hereafter are therefore an upper limit; this should be borne in mind if such numbers appear critical in decision-making. Conversely, the belief that there are levels of exposure below which no adverse effects are incurred is nowadays hard to sustain (Henderson, 1987).

5.4.2 Cancers

In the first stage of a cancer, the disturbance of body cells can create latent responses: some of the damaged cells can exist, for rarely less than three or four years and usually for decades, with little or no noticeable effects on an individual's health. At some stage, however, uncontrolled multiplication of damaged cells may occur, resulting in illness and often death. (An excellent account of the initiation of such cancers is given in Fremlin (1985, Appendix 1).) Some forms of such cancers can be initiated by radioactivity, including natural radiation from the ground and cosmic rays; cancers are also known to be caused by many chemical substances, some of which have been mentioned in Section 5.3.3. The chance that a cancer will develop after irradiation depends on the dose (or energy deposited) and its intensity and location within the body.

Body organs respond with differing degrees of sensitivity and according to the type of radiation; much research has been directed into determining weighting factors. Indeed such research is considerably more extensive than that undertaken on non-radiation carcinogenic effects with a correspondingly much greater understanding (Bengtsson, 1988). Data has been mainly derived from large-dose cases such as the Japanese atomic bomb victims or patients given heavy doses of X-rays several decades ago.

The scientific unit in current use for dose measurements is the sievert (Sv). This is related to the sum of effects on various parts of the body, each effect being dependent on the amount of energy deposited, its quality (alphas are more damaging than betas, for example), and the sensitivity of the given part to such damage. A dose of one milli-Sv (mSv) or 0.001 Svs, is then assumed (by the linear hypothesis) to correspond to an increased risk of cancer of 1 in 100,000. The sievert is 100 times larger than the previous dose unit, the rem. However, a more meaningful expression which relates dose to a convenient everyday experience, that of smoking, will be defined. A similar linear relation is assumed to hold for the effects of smoking cigarettes: in fact, smoking 150 cigarettes is estimated to incur a chance of 1 in 10,000 of developing cancer. It follows that reducing the figure to 15 gives a chance of only 1 in 100,000. Using an equivalence of 15,000 cigarettes for 1 sievert enables an 'everyday' appreciation of the valuations of hazards as listed in the table below, derived from data presented by Fremlin (1985). Here, the unit 'cig' refers to the effect of smoking one cigarette: for example, smoking 4 cigarettes in 1986, followed by another 8 cigarettes in 1987, would have roughly the same risk of cancer as one year's average dose from radon within a house. This data provides a useful background against which to judge the effects of power systems, to be quantified in Chapter Six.

Table 5.1

Annual Doses from Some Radiation Sources

Source	Annual Dose (cigs)
Cosmic rays	4.5
Potassium-40 within the body	2.7
Natural alphas within the body	2.4
External gammas	5.4
Radon within houses	12.0
Bomb test fall-out	0.2

It is interesting to observe that the largest dose in the above table, due to radon emanating into buildings, is to a large extent avoidable, i.e. the problem can be 'designed out' in new houses and surfaces sealed in existing houses. The cost of such remedial action would be far less than 'Rolls Royce' nuclear waste schemes and save many times more lives. This issue, though very important nationally, is, however, not relevant in discussions of the choice of power system.

Another useful yardstick is the Annual Limit of Intake (ALI), for which authoritative data is available in the literature (ICRP, 1981 and 1986). As its name implies, this defines, for a given radionuclide, its limiting acceptable level of annual intake into the human body, corresponding to a dose of 5 milliSieverts (mSv) or 75 cigs. From the discussion above, the number of cancers arising is proportional to the total number of ALIs received; the latter can therefore be a further useful indication of the hazard arising from the release to the biosphere of the given radionuclide. According to the scale of the dose received by large populations, the effects can be expressed as convenient, either as cigs, ALIs or probable number of cancers. It is perhaps worth emphasising here, that exceeding a radiation dose limit does not imply a dangerous situation - 2 ALIs in one year is merely equivalent to having smoked 150 cigarettes, i.e. a chance of 1 in 10,000 of developing cancer, a risk taken with equanimity by many smokers.

By way of example, the radioactive content of radium-226 in the oceans was deduced by the procedures of Appendix 1.2 to be 6 million TBqs (160 million Ci). The number of Bqs corresponding to an ALI for radium-226 is 20,000 (ICRP, 1981), so the toxic potential, i.e. if all the ocean radium-226 was ingested, is 300 million million ALIs: equivalent figures are 23,000 million million cigs or 15,000 million cancers. Another example is the discharge of uranium from spoil heaps and stacks from coal-fired power. Table 3.2 gives an emission of 5 tonnes per GW(E)yr. Corresponding to this, the total chain activity is about 0.3 TBqs (8 Ci) of alphas and 0.2 TBqs (5 Ci) of betas per GW(E)yr, equivalent, if all ingested, to 0.4 million ALIs, 34 million cigs or 23 cancers.

The above examples give useful bases for comparing in later chapters some of the potential effects of discharges from power systems.

Non-radioactive substances can create cancers through chemical effects, but it is often difficult to distinguish the various mechanisms because of the complex environments to which the victim has been exposed. However, using a similar proportional assumption as for radioactive effects above, small-dose effects can be predicted from the more serious case histories. A further complication is that different agents may interact, in some cases causing more damage than predictable by simple addition; in other cases, the reverse effect actually occurs (Bengtsson, 1988).

With both radiation and a large number of chemical carcinogens, genetic responses have been observed in experiments on mice, fruitflies, viruses, etc. (Cohen, 1983). Corresponding effects among human beings have been difficult to quantify, though they are certainly much less than the direct effects, as observed from survivors of the Hiroshima and Nagasaki atom bombs (Jones, 1987). In fact, the rates of abnormalities in babies conceived after the above bombing was no greater in these two cities than in corresponding Japanese cities unaffected by the bombs. Follow-up studies have confirmed there has been no increase in hereditary effects in Hiroshima and Nagasaki (Henderson, 1987). There is practically no quantitative information on the genetic effects of chemicals on human beings but there is no reason to believe that they are less from coal-fired power than from radionuclides produced by nuclear power.

5.4.3 **Non-Cancerous Diseases**

Nearly all non-cancerous detrimental effects of power production arise from the inhalation of polluted air. Diseases such as bronchitis can create chronic ill-health, which may result in death directly or reduce the resistance of the body to other diseases: the latter may then be recorded as causing death. A further complication in relating pollution to lung diseases is that the final years of life may be spent at a different location (even a health resort), so that the correlation of pollution and associated deaths geographically can be very difficult. A probable mechanism for lung and bronchial damage is associated with the smaller particles in stack effluents, which can more easily pass the filters. These have a high surface to volume ratio because of their small size. Consequently, toxic components which are volatile at stack temperatures can condense on their surfaces in high concentration on cooling. The particles can then be deposited as 'hotspots' in the lungs, as mentioned earlier for sulphur dioxide. Non-respiratory diseases can also be caused by pollutants. In addition to being carcinogenic, cadmium can cause damage to the liver and kidneys and is implicated in some forms of heart disease: it can be assimilated from smoke particles (Price, 1983). Mercury can be passed in a food chain to fish which then becomes dangerous for human consumption (Price, 1983). Like mercury, lead can cause brain damage: though small concentrations are unlikely to cause death, the cumulative effects on large populations, particularly children, can be serious (Greenhalgh, 1980).

5.5 **SUMMARY**

Of the various forms of nuclear radiation, alpha radiation is mainly hazardous to the human body internally after ingestion or inhalation: it can provide a highly localised dose. On the other hand, beta and gamma radiation cause less concentrated damage: the latter in particular can penetrate the body deeply from outside. Examples of these effects arise from buried nuclear wastes and the uranium and thorium decay chains present in coal-fired power stack emissions and leachates from coal spoil heaps and fly ash lagoons. There are several natural sources of radiation which between them strike each human cell several times every year.

Chemical pollutants from power systems arise almost exclusively from coal combustion. Trace concentrations of organics and metallic elements, some with carcinogenic properties, are discharged in stack gases in tonnage quantities per GW(E)yr. From the dumps of fly ash and coal spoil, many thousands of tonnes of chemicals are leached into natural water flows, eventually to reach the sea. Huge tonnages of inorganic gases also pour out from the stacks and, again, many tonnes of fine particulates evade the filters, thus creating inhalation hazards. New combustion techniques can only provide a small reduction in acid gas emissions; even these create huge disposal problems and may merely change the route by which toxicants enter the environment.

Cancers can arise from both radiation and chemical effects. Knowledge of the latter is relatively poor: even those with anti-nuclear views admit that much more is probably known about the consequences of exposure to radiation than most other environmental hazards (Patterson, in Foley, 1978). Although there is no clear evidence that the detriment to man is proportional to dose, i.e. there is no threshold below which there is zero effect, a pessimistic approach is usually assumed, i.e. that a small dose gives a correspondingly small chance of cancer initiation. Genetic effects of radiation appear to be insignificant (Section 5.4.2).

Respiratory diseases are probably the main immediate detriment from coal power stack discharges (though the long-term effect of huge quantities of chemicals spewed into the biosphere may also be serious). A plausible explanation is that the highly concentrated condensation of acid gases on fine particulates overwhelms body defences when the 'fines' are deposited in lungs. It has been assumed in the past that there is a threshold concentration below which the detriment can be neglected, though nowadays there is uncertainty on this point (Henderson, 1987).

The oxides of sulphur and nitrogen are the principal sources of Acid Rain, which causes widespread damage, both directly in the atmosphere and indirectly through chemical reactions after deposition, releasing more toxic pollutants, e.g. aluminium. Carbon dioxide, the chief product of coal combustion, may eventually have even more serious effects, through the absorption of infra-red radiation emitted from the Earth's surface - the Greenhouse effect.

Overall, discharges from coal-fired power stacks are so huge and complex that it can be difficult to pinpoint a constituent responsible for a particular detriment. On the other hand, the detection and identification of various types of radiation from nuclear power discharges is possible with extremely sensitive instruments to very low concentrations indeed: curiously, however, the advantages over fossil fuel emissions of such sensitivity has become almost a penalty in the public perception of nuclear power, since many people confuse detectability with hazard.

Chapter Six

CASUALTIES FROM ROUTINE OPERATIONS

6.1 GENERAL

In Chapters One to Four, the various processes for UK electricity production have been outlined, with requirements for normal construction, operation and decommissioning. Unfortunately, both operations on process sites and the associated discharges may involve unplanned effects, incurring injury or death to both operators and the general public. In this chapter, we set out to quantify these undesirable side-effects of normal operations and also accidents to individuals or small groups. Accidents involving larger groups of people are discussed in Chapter Seven.

Before investigating the above effects, it is worthwhile reminding ourselves of the overall objective, i.e. to compare the various possible UK power systems. In order to do this fairly, each system must be assessed with the same ground rules. An important aspect of these is the period over which any detrimental effects should be evaluated. For example, pollution from stack discharges from coal-fired power stations is conventionally assessed as the inhalation damage, eventually appearing as bronchitis, cancers, etc., inflicted before the pollutants have been deposited on the ground. On the other hand, the safety of nuclear waste is often queried in relation to very long-term changes, say the migration of neptunium into drinking water after many thousands of years. An obvious question which then springs to mind is whether the same diseases will still be prevalent in the distant future. Cohen (1983) has analysed this point and suggested that it is likely that the various forms of cancer will be of little importance in 500 years' time. Even so, he presents figures for effects up to 500 years and for much longer time periods as if no progress over the present ways of treating cancers will be made. We will follow the same approach in the analysis below in order to test whether our system preferences are dependent on improvements in the future treatment of the appropriate diseases: the 500-year figure will normally be quoted, with an additional 'long-term' figure, covering thousands of years, where necessary.

6.2 NUCLEAR POWER

6.2.1 Construction

Building nuclear plants, either for power or for fuel processing, involves few hazards novel to the construction industry, since, before the plants are commissioned, there are virtually no radioactive operations. From statistics such as those given by Greenhalgh (1980) and Inhaber (1982), estimates of death rates during the construction of both power and fuel cycle plants, averaged over an assumed plant life of 30 years, are equivalent to about 0.01 deaths per GW(E)yr.

6.2.2 Fuel Preparation

It is debatable whether uranium mining accidents are relevant to UK nuclear power production, since all the mining takes place overseas. Nevertheless, it is interesting to check the corresponding rate. In mining uranium, there are two main hazards: first, physical accidents and second, lung disease due to the inhalation of radon and its decay products. The second of these, according to Beckmann (1976), results in about 0.02 deaths per GW(E)yr: this figure includes a small contribution from subsequent milling of the ore. Deaths from physical accidents were less by a factor of 10. In recent years, improvements in ventilation have reduced radon concentrations twentyfold. Reasonable death rates for future mining accidents and lung disease would therefore be about 0.002 and 0.001 per GW(E)yr respectively.

In the manufacture of uranium fuels there is relatively little radiation hazard involved. The radiotoxicity of refined uranium, as mentioned in

Chapter One, is less than its chemical toxicity, which itself is less than that of materials used currently in many non-radioactive processes in the UK.

In the preparation of plutonium fuels, there is, in principle, the possibility of a 'criticality' accident. This type of accident can happen when plutonium occurs in a combination of mass, concentration and shape sufficient to sustain a chain reaction of fissions. Unlike a reactor, where this reaction is controlled, a sudden burst of energy is produced, as in a 'conventional' chemical explosion. There is no possibility of an atomic explosion (see Section A.1.4.2 of Appendix 1), but there may be a serious hazard to nearby operators, mainly from neutrons and gammas. In many instances, the chance of a criticality accident can be eliminated by design, e.g. by limiting the diameter of columns containing liquids so that criticality is precluded by shape. Further, after dilution of the plutonium to about 20% in uranium as required for Fast Reactor fuels, the hazard is greatly reduced. The fabrication of Fast Reactor fuel elements, which currently occurs mainly at Sellafield, presents a potential hazard from inhalation of dust containing plutonium; the standard of operator protection must therefore be high. To date, contamination of operators has been very infrequent, with very few serious casualties. There are no reported effects on the public.

Transport of new fuels to reactor sites involves relatively few journeys; furthermore, there has been no significant spread of radioactivity in the few accidents that have occurred, and so negligible impact on the public.

Overall, therefore, the only significant detriments recorded to date and likely in the future in fuel preparation arise during the mining of ore at the rates evaluated above.

6.2.3 Power Operation

Accidents not associated with radiation occur of course on nuclear power stations, but, because of the relatively few staff involved, death rates are small - say at least as low as for coal-fired stations, derived later as 0.1 deaths per GW(E)yr. Corresponding figures due to radiation can be deduced from Cottrell (1981) to be about 0.01 deaths per GW(E)yr. For the general public, data in UNSCEAR (1977) for nuclear station stack discharges suggest that deaths per GW(E) yr would be 0.02 from krypton and xenon isotopes and 0.01 from tritium.

Though there have been suggestions of cancer clusters near nuclear plants, the numbers involved, even if substantiated, would only be about 0.02 per GW(E)yr (see Appendix 6.1).

6.2.4 Reprocessing Including Current Disposal of Wastes

Transport of spent fuel from reactors for reprocessing in the UK, mainly at Sellafield, has proceeded for many years with no serious accidents; this is true, too, of such transport overseas. Even so, public anxiety has been engendered by the media, so that in both the UK and the US realistic demonstrations of extreme accident conditions have been carried out with spent fuel flasks on trucks drawn by locomotives. In spite of attempts by Greenpeace to decry the realism of such demonstrations, there appears to have been substantial allayment of the public's fears.

With respect to reprocessing operations on Sellafield site, surveys of workers have shown that there is a lower incidence rate of cancers than among the general UK population. (This result, known as the 'healthy worker' syndrome, is common among large workforces, since some less healthy applicants are screened out during recruitment.) If, therefore, any cancers have been caused by process operation, the number must be very small. From estimates of the total dose to the workforce, a possible figure is one per year, probably more than compensated by the special health care on the site. With tighter regulations on permissible doses, a fivefold reduction of this figure in the future is probable. At Dounreay, there is a smaller workforce than at Sellafield; it is not surprising, therefore, that radiation accident figures are relatively small. Non-radiation accidents for all UK reprocessing result in less than 3 deaths per year - say about 0.6 deaths per GW(E)yr. It would appear fair to assume that (say) a fourfold increase in fuel throughput in future larger plants will require perhaps only a doubling of operating staff. An overall level of 0.3 deaths per GW(E)yr might then result, most of which would not be caused by radiation.

From the reviews in earlier chapters, the main hazards to the general public in the UK, from routine discharges during current reprocessing operations, arise through ingestion or inhalation of radioactivity by routes described below.

A. The main discharge of activity to sea from nuclear sites occurs at Sellafield. However, the more important ingestion routes, from eating fish of various kinds caught within a few miles of the end of the pipeline, have been monitored carefully over many years: the corresponding doses have been kept well within internationally agreed limits. The prediction of low impact locally is not surprising, considering the vast dilution afforded by the sea and the distances fish can roam before they are caught. Fremlin (1985) estimates that reprocessing at Sellafield associated with power production up to 1980 would probably have caused about 0.2 deaths per GW(E)yr, commenting, however, that future fuels from AGRs and PWRs would be cooled for longer periods before reprocessing, thus reducing nuclide activities and associated discharges still further. Clarke (1982) estimates the dose from Sellafield pipeline discharges in 1978 to be 0.13 Svs (13 rems). This corresponds to a probable number of cancers of 1.3 or about 0.3 cancers per GW(E)yr, in good agreement with Fremlin's figure.

About 7% of the plutonium annually put down the pipeline enters the Atlantic through the North Channel of the Irish Sea. The sea-bed sediments provide a 'sink' removing much of the plutonium from solution. However, some of the sediments may then return to land; it is estimated that such a return amounts to perhaps 0.01% of the total discharges to date (RWMAC, 1985). Of this, there is rather more returned to estuaries than to beaches. The effect is reversible, at least to some extent, since there is a reduction of activity in the Ravenglass Estuary following a lower discharge from Sellafield. In the future, there should be much less activity in the sea, because of the recent installation of extra purification plant costing several hundred million pounds, including construction, operation and decommissioning costs. The probable saving of life, however, even over many years, is likely to be small.

At Dounreay, discharges into the sea and atmosphere have been much lower than at Sellafield and the population near the site is much smaller. Levels of activity in the water draining from the site waste pits are trivial, with negligible consequent discharge to sea.

Overall, it would seem reasonable to use the above estimates for Sellafield pipeline discharges when treating future AGR and PWR fuels to give a figure of about 0.01 deaths per GW(E)yr in the general public as a consequence of UK reprocessing.

B. The most important discharges to atmosphere from reprocessing sites are also at Sellafield. As a result, the public can receive activity both directly by inhalation and indirectly from deposits on the ground, where, for example, contaminated grass may be eaten by cows and the resulting milk drunk by children. (Active spray from sea discharges can be blown on to land, thus contributing to this method of contamination.) The reprocessing plants are designed, however, with all these pathways to man in mind; very low doses to the public are assured by observing stringent regulatory limits for stack releases. Estimates by Sir Edward Pochin for the Windscale Enquiry (Parker, 1978), concerning atmospheric discharges of tritium, carbon-14 and krypton-85, indicated that the effect of the new THORP reprocessing plant, supporting about 20 GW(E) of reactors, might be about 2 cancers per year - say 0.05 deaths per GW(E)yr; this would be halved if krypton-85 were removed from the stack gases. Levels from Dounreay stacks are much lower than at Sellafield.

C. LLW, disposed by the current practice of shallow land burial, can in principle be leached by water and eventually ingested by drinking, eating plants or consuming animal products. At Drigg, waste buried in shallow trenches can be contacted by rain percolating through the simple soil cover. However, the migration of activity is so slow that water draining to the adjacent stream, closely linked to the Ravenglass Estuary, would be satifactory for drinking by man (BNFL, 1986) - if the local cattle were not already imbibing there! Wastes with leach water effectively draining to sea at Dounreay have even lower potential effect - the estimated dilution in (undrinkable) local seawater is estimated to be of the order of one million billions! It would seem fair, therefore, to ignore casualties from disposal of LLW.

6.2.5 Future Disposal of LLW and ILW

By the end of the century, NIREX proposes to have a new LLW disposal facility operating. This was originally proposed as burial in trenches as at Drigg and Dounreay, but wastes would be packed in drums and placed inside concrete cells before being cemented over and covered with a clay cap (see Section 4.1.7). There would be a negligible content of long-lived radionuclides, so that, assuming the natural clay strata prevented migration of any leached activity for a few hundred years, there would be no significant hazard to the public at any time. It would seem likely that such retention of activity will occur. There is an inherent difficulty, nevertheless, in verifying that no leakage from the (waterlogged) waste has occurred, and, if it has, in assessing how to carry out satisfactory remedial action. In other words, the concept suffers from the same possible faults which have caused difficulties on containment-type sites in the US.

However, it was recently announced (May 1987) that the chosen four trench sites have been abandoned, the costs being not much less than the incremental costs of burying LLW in a deep burial system necessary for ILW. Migration of activity from these should be even lower than from the above designs of trench.

The volume of the ILW, mostly from reprocessing, is much smaller than that of the LLW discussed above. There is therefore an option to store these for quite long periods before disposal by deep land burial.

Since it is shown in the following sections that the hazards from HEW and spent fuel after deep burial are less than those from coal-fired power discharges, it is reasonable to infer that ILW similarly buried would cause a negligible number of deaths.

6.2.6 Future Disposal of Spent Fuel and HEW

6.2.6.1 Potential Hazards relative to Natural Conditions. Previous sections have discussed the possible impact on critical groups who receive the highest individual doses from a radioactive release, probably by living or working near the discharge point. Very slow releases, such as migration from a repository through thick sections of rock, cause negligible local hazards, but the sum of very small effects to members of large populations (i.e. the collective dose), can result in a certain number of cancers (Section 5.4.1). Using similar evaluations in Appendix 6.2 to those made for natural radionuclides in Appendix 1.2, it is found that, considering nuclides of appreciable activity after a few hundred years storage,

(a) spent fuel, equivalent to forty times the total world generation to date, would only have the same level of radioactivity as that naturally present in the oceans, if similarly dispersed in solution,

(b) spent fuel of similar age and equivalent to world output to date, would only have a quarter of the ingestion toxic potential (Section 5.4) of the natural uranium-238 and its decay products in the oceans, if it were similarly dispersed in solution in the oceans, and

(c) if the most important contribution to the radioactivity and toxic potential, americium-241 (half-life 433 years), could be withheld from entering the sea for one or two thousand years, the long-lived radioactivity and toxicity of spent fuel as in (a) and (b) would be decreased severalfold.

It can be concluded that the slow release of nuclear wastes to sea will create no substantial increase in seawater radioactivity or toxicity, either locally or globally, if the leakage does not occur for a few hundred years, and that appreciable advantages are gained if americium-241 can be contained for one or two thousand years. These effects are also improved considerably if plutonium is virtually absent, as in HEW.

6.2.6.2 Potential Hazards Relative to Coal Power Discharges. A second yardstick against which the hazards of nuclear wastes can be measured is the currently accepted hazards from coal-fired wastes. Calculations in Appendix 6.2 compare the hazards from coal-fired power stacks with those from leakages to sea of various nuclides in spent fuel and HEW. Similar conclusions may be drawn as in the comparison with natural radioactivity in the oceans, i.e. it is important to prevent significant leakage from the above nuclear wastes for a few hundred years and desirable to contain americium-241, say for one or two thousand years. The dispersal of neptunium-237 in the sea at any time is not important.

6.2.6.3. Assessment of Nuclide Migration. A third approach to the safety assessment after waste burial is to calculate the changes taking place thermally and in the movement of nuclides in surrounding rock. The characteristics of heat emission during burial of nuclear wastes have been described in Sections 4.1.8.2 to 4.1.8.5 and Appendices 4.1 and 4.2. If the associated problems can be avoided by careful design, then the remaining hazard after disposal depends on the migration behaviour of the long-lived nuclides, key ones being neptunium-237 (half-life 2.1 million years), plutonium-239 (half-life 24,400 years) and americium-241 (half-life 433 years), as mentioned above. Though there are many nuclides, e.g. strontium-90 and caesium-137, which may have a much higher activity in HEW or spent fuel at the time of disposal, their half-lives are much shorter. If, therefore, the above key nuclides can be shown not to be hazardous, it is usually easy to infer that other nuclides can be ignored. Detailed calculations, e.g. Burton (1981) and Hill (1978, 1980), bear out the validity of this useful simplification.

The calculations discussed in Appendix 4.4 show that, given certain assumptions on the properties of the rock, groundwater flow and nuclides, the fraction of a particular nuclide which escapes after burial at a practicable depth is insignificant. However, in order to allow for possible shortcomings in the rock, the general philosophy, accepted in most countries, is to provide extra barriers to nuclide movement (see Section 4.1.2). These can be the waste form itself, e.g. glass or the synthetic mineral 'SYNROC' (see Appendix 3.1), or thick sections of metal and/or clay round waste packages (see Fig. 4.9(c)). On the other hand, if a disposal location under the sea-bed (but for convenience accessed from land) is chosen, the absence of significant water flow (see Appendix 4.3) and the relative harmlessness of radionuclide migration into saline rather than freshwater, would make the above extra (expensive) man-made barriers seem unnecessary.

The above discussion outlines the principles on which many concepts have been developed for the deep burial of HLW, concepts both practicable and safe, but considerably overdesigned in relation to coal power and other toxic discharges. The levels of plutonium assumed in the waste, however, are relatively small, say 0.1% of that in spent fuel. In the case of disposal of the latter, proof of adequate safety after disposal is virtually determined by the behaviour of plutonium. Moreover, it is not practicable to alter the spent fuel itself; consequently, external man-made barriers and superior natural barriers must be considered. A Dry Repository with natural draught cooling for as long as deemed necessary could be attractive here, since the emplacement tunnels are wide enough to allow substantial external barriers to be installed. At the same time, the possibility of retrieval for reprocessing would be maintained. On the other hand, the relatively small diameter of fuel assemblies lends itself to disposal down holes of narrow bore, which themselves are easier to drill to greater depths than (say) one-metre diameter holes necessary for glass block disposal. Some simple calculations in Appendix 4.4 suggest that such disposal under the sea-bed could be made acceptably safe. Disposal of long-lived active wastes, such as PCM or ILW, could similarly be at considerable depth as HEW but, with lower heat output and external radiation levels, emplacement would be easier than HEW.

Overall, it is reasonable to deduce that there are a number of storage/disposal arrangements for HEW and spent fuel which are acceptable, especially in relation to the disposal of many toxic wastes today and natural radioactive materials (see Appendix 6.2), and, in particular, those discharged from coal-fired power operations, to be discussed in Section 6.3. The use of locations near the coast, either under the sea-bed or as a Dry Repository in a coastal hillside, can reduce uncertainties in assumptions underlying the safety calculations. Corresponding hazards to the operators and the public are negligible.

6.2.7 **Decommissioning**

The bulk of nuclear power plant decommissioning is similar to non-radioactive demolition operations with additional appropriate precautions such as the prevention of inhalation of contaminated materials. Since these operations will be only a tiny fraction of the total demolition work in the UK, it may be deduced from the corresponding national statistics that the number of deaths will be very small.

The more active sections from reactor cores, requiring cutting up, packaging and disposal, will involve more sophisticated handling. However, from the experience in dismantling and refurbishing the D1206 plant at Dounreay (see Section 3.2.5), it would not be expected that death rates would be significant.

6.3 POWER FROM COAL

6.3.1 Construction

Sinking shafts for new mines is a hazardous business for the individuals involved: both this type of construction and that of building coal-fired power stations, however, take a relatively short time and involve a small number of men compared with those required in the eventual extraction of coal. Consequently, the numbers of casualties are by comparison slight: a figure of 0.01 deaths per GW(E)yr is estimated here, based on a general survey of engineering construction industries.

6.3.2 Preparation of Fuel

Statistics of small-scale mining accidents in successive recent annual reports of British Coal, e.g. NCB (1985) may be scaled per GW(E)yr to give 0.5 deaths, 10 serious injuries and 200 less serious injuries requiring an absence of over 3 days from work. Extrapolations from US figures (Beckmann, 1976) give good agreement with the above, whereas Greenhalgh (1980) suggests twice as many deaths. Deaths due to lung diseases from mining are more difficult to assess, since they can occur decades after symptoms begin. The quality of life can be reduced for a long time (as discussed in Chapter Eight) and death can occur apparently through other causes, though with a contributory factor from lung problems. Fremlin (1985) suggests there may be about half as many deaths from lung diseases as from physical accidents: death rates given by Beckmann (1976) appear to be much higher. The rate of certification of lung disease in the UK a few years ago was about ten times the death rate (HSC, 1978). Overall, it would seem fair to use a figure of 2 deaths per GW(E)yr resulting from the extraction of fuel for use in coal-fired power stations.

The transfer of 3.5 million tonnes of coal per GW(E)yr of power requires a considerable flow of trains or heavy goods vehicles. Accident rates can be deduced from road oil tanker statistics (HSC, 1978) to be about 0.05 deaths per GW(E)yr. From coal train accident data (HSC, 1978), rail deaths at a similar level may be inferred.

6.3.3 Power Operation

The number of operators in coal-fired power stations is small and the jobs are safe relative to the mining workforce, so that it is not surprising to find an estimate of only 0.1 deaths per GW(E)yr (HSC, 1978).

The effect of coal-fired power on the public is discussed in Chapter Eight of Fremlin (1985): from an analysis of bronchitis statistics in the late 1940s related to the much higher death rate in Britain than the Continent, it was deduced that between 20,000 and 60,000 excess deaths from bronchitis occurred in British cities each year. Open domestic fires no doubt provided an important fraction of these and, since the 1952 London 'smog' disaster, regulations have reduced these discharges of smoke considerably. On the other hand, many deaths recorded as heart failure could have bronchial problems as the primary cause.

Modern designs of power station, with high stacks and possible future trapping systems for sulphur dioxide, could reduce hazards from the latter, but the higher temperatures of combustion produce more nitrogen oxides. Moreover, both the latter and toxic elements in coal - arsenic, beryllium, cadmium, lead, thallium, thorium and uranium - some of which are carcinogenic, volatile in stack gases and not taken out either currently by filters or in future trapping systems, can condense on particles and settle on lung cells as described in Sections 5.3.3 and 5.3.4. Because of the complex conditions and resulting uncertainties, many authors quote a range of respiratory death rates per GW(E)yr from stack discharges, e.g. 40 to 100 (Forbes, 1974), 20 to 60 (Greenhalgh, 1980), 10 to 200 (Hamilton, 1977), 20 to 70 (Inhaber, 1982), 22 (Novegno, 1987), 45 (Pochin, 1976) and 40 to 100 (Wilson, 1974).

Though several of these references apply to US conditions, corresponding figures for the UK, with its higher average population density, are unlikely to be lower. It would not seem unreasonable, therefore, to use the rough average of 50 in subsequent comparisons of systems hereafter. This does not take account of chronic respiratory diseases, asthma attacks, aggravated heart-lung diseases, etc., which for the US can range per GW(E)yr from about 200,000 (remote location) to about 600,000 (urban) (Comar, 1976): nor are the indirect consequences of Acid Rain specifically included, such as from increased aluminium and lead in drinking water.

The excess cancer deaths caused by some of the above inorganic toxicants and hydrocarbons such as benz-alpha-pyrene (BAP), discharged in tonnage quantities annually from stacks of coal-fired power stations, are difficult to estimate, partly because of the latent period of decades before the appearance of symptoms, but also from the similar and concurrent effects from domestic fires, car exhausts, etc. The Commission on Energy and the Environment (1981) suggests that perhaps 150 cancer deaths per year arise from air pollution, of which perhaps one-third to one-half are due to power stations, i.e. 1 to 1.5 per GW(E)yr. Fremlin's estimate of about one extra cancer death per GW(E)yr from this source agrees satisfactorily with this figure. Camplin (1982) estimates that radioactivity in coal-fired power stacks causes about 0.1 deaths per GW(E)yr, several times the rate from nuclear power stacks (see Section 6.2.3). Cohen's estimates are similar at the 500-year period; of considerable interest is that his figures for chemical carcinogens and radon rise to 70 and 30 respectively in the long term. This is an *important point*, in that, if long-term cancer hazards are believed to be important, they are hundreds of times more serious from coal-fired power stacks than from nuclear wastes and discharges. On the other hand, if only short-term hazards matter, there are no problems in the disposal of nuclear waste: simple 'dry' trench burial would be adequate.

It is difficult to estimate the hazards from the huge spoil and fly ash dumps associated with coal-fired power. For each GW(E) of power, the annual dumping of 1 million tonnes of spoil and 0.3 million tonnes of fly ash would contain large quantities of toxic and radioactive chemicals (see Table 3.2). The major hazard source here is from leaching and consequent pollution of surface waters and groundwaters (Grove-White, 1985). The Health and Safety Commission (1978) accepts that, for each tonne of fly ash, there is 1 gramme of toxic and radioactive material in a leachable form, so that eventually large quantities migrate via the above waters to the sea. Again, using Cohen's long-term estimates above, the hazard will be much greater than from the corresponding amount of nuclear waste buried beneath the sea-bed. On the other hand, it is claimed (HSC, 1978) that only in a few cases do concentrations exceed safe levels for drinking and that existing treatment methods ensure that none reaches drinking water supplies. If this approach is regarded as satisfactory, it is difficult to see how burial of wastes as currently carried out even in the present relatively rudimentary way at Drigg and Dounreay could be described as unsatisfactory.

6.3.4 Decommissioning

The demolition of coal-fired power plants is only a small fraction of general demolition work in the UK, and it would seem reasonable to deduce that corresponding casualties are negligible. A more serious problem is the flooding of abandoned mines, referred to in Section 3.3.3. Though polluted waters are probably of little danger to man, wildlife can be seriously affected. The River Girvan in Ayrshire was polluted to such an extent from a nearby closed mine that all fish (including a prime salmon fishery) were killed off (Comm. En. Env., 1981). However, it is difficult to quantify the effects of such flooding and subsidence in general and these are not listed hereafter.

6.4 FUSION AND RENEWABLE POWER SYSTEMS

6.4.1 Construction

Fusion power should not cause a significant number of casualties during construction - say 0.01 deaths per GW(E)yr by comparison with nuclear power. Tidal power schemes involve substantial construction work. The installation of wave and wind machines is also relatively dangerous; from a comparison of erection worker casualties with those of general building workers, a nominal level of one death per GW(E)yr has been derived for each of these systems.

6.4.2 Fuel Preparation

The only process here requiring fuel is fusion power, which for each GW(E) needs a few tonnes of heavy water and probably lithium per year. The lithium supply is small and casualties during production would be slight. It can be deduced from the heavy water plants supplying Canadian fission reactors that the number of deaths corresponding to 1 GW(E)yr of fusion power would also be negligible.

6.4.3 Power Operation

Smaller routine casualty rates would probably arise with fusion power than in fission power during electricity production because of the simpler processes, even allowing for the arisings of 2,000 tonnes ILW/GW(E)yr (Table 3.3). Routine discharges of tritium would only be above a tenth of that from nuclear fission sites.

Tidal barrages require few operations during power production and therefore have negligible casualties. On the other hand, though there is little significant experience on which to assess hazards of wave power operation, routine accidents could be perhaps 1 death per GW(E)yr, since the structures are extensive and might have to be inspected in adverse conditions. For wind power, too, there has been little experience to date. By analogy with steeplejack accidents, the death rate could be relatively high - say 0.5 per GW(E)yr, unless, as in the vertical axis design, lowering of the blades to ground level for servicing were practical.

Though conservation has not been generally discussed in previous chapters, there have been some interesting observations on the buildup of radon in houses when ventilation is reduced to cut down heating costs (Cohen, 1983; Fremlin, 1985 and 1986). Radon is formed during the decay of radionuclides present in the ground and in building materials. It permeates the floors and walls to contaminate the air of rooms. Cancers can then result as described in Section 5.2. Perhaps 400 deaths occur in this way every year in the UK. The Royal Commission on Environmental Pollution acknowledges this hazard and recommends that increased insulation should not mean inadequate ventilation (RCEP, 1984). A recent NRPB report by Wrixon (1985) suggests that radon levels in some houses are sufficiently high to justify the setting of safety standards.

6.4.4 <u>Decommissioning</u>

By comparison with fission power, the decommissioning of fusion power plant, though also involving activated structural materials, would seem somewhat easier in the absence of fission products and heavy isotopes. Casualties would thus probably be negligible.

Removing wave and wind power units would seem to involve only standard operations of the demolition industry and it is reasonable to neglect the associated casualties. It would seem probable that tidal barrages would not be demolished.

6.5 SUMMARY

Below are listed the figures evaluated through this chapter for the various phases of power systems. Table 6.1 presents data for operators and Table 6.2 the corresponding figures for the general public. Where these have been summed, the terms '500 year' and 'long term' refer to the methodology of Cohen (1983), discussed in Section 6.1.

Table 6.1

Summary of Estimated 'Routine' Death Rates of Operators per GW(E)yr for Various Power Systems

System/Phase	Nuclear	Coal	Fusion	Tidal	Wave	Wind
Construction (a)	0.01	0.01	0.01	1	1	1
Fuel Preparation	vs	2.1	vs	-	-	-
Power Operation	0.1	0.1	0.1	-	1	0.5
Reprocessing	0.3	-	vs	-	-	-
Waste Disposal	vs	vs	vs	-	-	-
Decommissioning	vs	vs	vs	-	vs	vs
Total	0.4	2.2	0.1	1	2	1.5

Notes: vs denotes very small, i.e. less than 0.01.

(a) Construction and decommissioning figures are averaged over the plant lifetime.

Table 6.2

Summary of Estimated 'Routine' Death Rates for the General Public per GW(E)yr of Various Power Systems

System/Phase	Nuclear	Coal	Fusion	Tidal	Wave	Wind
Construction (a)	vs	vs	vs	vs	vs	vs
Fuel Preparation	vs	0.1	vs	-	-	-
Power Operation	0.03	(b)	vs	vs	vs	vs
Reprocessing	0.06	-	vs	-	-	-
Waste Disposal	(c)	vs	vs	-	-	-
Decommissioning	vs	vs	vs	vs	vs	vs
Total (500 year)	0.1	50	vs	vs	vs	vs
Total (Long term)	0.1	150	vs	vs	vs	vs

Notes: vs denotes very small.

(a) Construction and decommissioning figures are averaged over the plant lifetime.

(b) 500 year figures are 50 for respiratory diseases, 1 and 0.1 for cancers from stable and radioactive chemicals respectively; long term figures for cancers are correspondingly 70 and 30 (Cohen, 1983).

(c) The '500 year' and 'long term' figures are very small and 0.01 respectively.

Clearly, by far the greatest death rate arises from coal-fired power; the main short-term contributions here are from respiratory diseases resulting from stack discharges. Radioactivity in these discharges causes several times more cancers than for all phases of the nuclear system, and stable chemicals from coal create a further tenfold increase. In the long term, these cancer rates increase a hundredfold, whereas nuclear power cancers remain small, since nuclear wastes are carefully disposed away from the biosphere to preclude long-term hazards.

In drawing up figures for nuclear wastes, it has been shown above, and in earlier chapters, by comparison with similar effects naturally and from coal-fired power systems,

(a) that disposal of LLW in trenches will result in negligible hazard if the waste is always above the groundwater table,

(b) that deep burial of HLW will cause no serious hazard if release of nuclides is prevented for a few hundred years and thereafter only occurs into saline water,

(c) that a useful further reduction in hazard compared with (b) is obtained by preventing the release of americium-241, i.e. by containment for one to two thousand years, and

(d) the release of neptunium-237 from HLW under such conditions is unimportant.

A large increase in world nuclear power (say tenfold) would not invalidate these conclusions. The implications of the eventual release to the oceans of all the plutonium in spent fuel from such a programme operated on a Once-Through basis would, however, require careful assessment.

In general, fusion and renewable systems appear to hazard the public less than nuclear power, as would be expected, though occupational casualties can be greater.

Appendix 6.1

UNPLANNED NUCLEAR REPROCESSING EVENTS WITHOUT SERIOUS NUMBERS OF CASUALTIES

Some unplanned events occurring during reprocessing at Sellafield are reviewed in HSE (1981), Appendix 1. None of these resulted in serious numbers of casualties, so the more important of them are recounted here rather than in Chapter Seven.

In September 1973, an accumulation of fission products, composed mainly of noble metal elements not easily soluble in nitric acid, caused a pressure surge, which forced a 'puff' of radioactivity, chiefly from ruthenium-106, to escape through shaft seals of pumps into the air of operating areas. As a result, 35 employees were contaminated but no deaths have resulted: the plant has remained shutdown since that time, since its capacity was relatively small and AGR and PWR fuel could be stored safely to await the opening of the new THORP plant.

In December 1976, it was discovered that activity had leaked into the ground from Building B38, used for the underwater storage of fragments of Magnox cladding, which had been stripped off the fuel before dissolution of the latter in nitric acid. A small fraction of fuel normally adheres to the cladding and is therefore a source of activity in the silo. The leak was eventually traced and stopped.

In October 1978, leaks from silos containing radioactive liquor contaminated the ground near the disused Building B701 with about 5000 TBq of activity, mostly at a depth of about 4 metres. Interestingly enough, the official investigation of the incident (HSE, 1980) concluded that it was unlikely that there would be any danger to the public at any time in the future in view of the remedial work being carried out to inhibit the migration of activity. (This entails the injection of bentonite mud into the ground.) If this is indeed true, there would seem to be a very high degree of safety in injecting HAL several hundred metres underground, a possibility mentioned in Appendix 4.5.

In November 1983, a radioactive mixture of aqueous liquid, solvent and solids was discharged into the Irish Sea under weather conditions which brought about the return of some of the solvent and solids to the beach. The latter was sufficiently contaminated as to be closed to the public for several weeks. Though there was no serious hazard to operators or the public, this spectacular 'own goal' by BNFL led to more restrictive regulations on discharges, as discussed in Section 4.1.3. At about the same time, an apparent 'cluster' of leukaemia was discovered near Seascale, a few miles south of Sellafield. The concentration of these cases was several times the national average, though the actual number of cases of excess deaths in Seascale ward, four, was small. The result of the Enquiry (Black, 1984) into the causes of the cluster was inconclusive, partly because of the difficulty of interpreting the statistics for such a small number of cases and the fact that other apparent UK clusters occur away from nuclear sites (Forman, 1987). More recently, a search of Sellafield records has shown that a release of 20 kilogrammes of uranium to sea occurred in 1957. This fact was not known to the Enquiry, which had been given a level of release of 0.5 kilogrammes of uranium. The change in release figures might therefore seem a possible explanation for the Seascale cancer cluster; however, each square kilometre of sea off Seascale contains tens of kilograms of natural uranium, so this explanation appears doubtful.

Craft (1987) gives a well-balanced account of the Seascale leukaemia problem, providing statistical evidence of the existence of a cluster among children there, but also citing other clusters well away from nuclear installations. Overall, Craft concludes that the Black Report reached the best interpretation of the data available at that time and that the 'Sellafield effect' is not proven. Subsequently, Gardner (1987) revealed unambiguous evidence from samples of 1,000 children born in Seascale and 1,500 children born outside the parish but attending local schools, that the leukaemia effect occurs solely in the first group and not at all in the second group. Even more recently, Openshaw (1988), in a careful and sophisticated study of leukaemia in Northern England, not only confirmed the Seascale cluster, but also found a much bigger cluster on Tyneside, over 35 miles from any nuclear installation. The final observations of Openshaw are important. 'The present analysis gives considerable weight to the hypothesis that there are likely to be environmental, perhaps pollution-related, causes of leukaemia other than radiation.'

In June 1988, the Committee on Medical Aspects of Radiation in the Environment, COMARE, set up in 1985 following a recommendation of the Black Enquiry, reported on the higher level of childhood leukaemia near Dounreay (six cases between 1979 and 1984 instead of one, expected on the national average). It suggested that the evidence supported the hypothesis that some (unknown) features of the Dounreay works increased the risk. However, they also ordered an inquiry into the possibility of radio waves, from the nearby US naval base, causing the leukaemia. Further, COMARE evaluated the effects of the 1957 uranium discharges at Sellafield, but concluded that they did not account for the excess cases observed.

The above leukaemia discoveries and studies have been reviewed by Taylor (1988). It is there pointed out:

(a) there were no excess leukaemia cases in other towns and villages near Seascale,

(b) there was no correlation between incidence of leukaemia with variations in discharge levels, and

(c) other clusters occur either nowhere near nuclear installations or near such installations with much lower discharges than Sellafield.

The review shows how sensitive clusters are to the choice of boundaries, and shows that radiation was unlikely to be the cause of the Dounreay cluster. The authors emphasise the need for noting the distribution of both increases and decreases in interpreting results and draw attention to the careful approach of Openshaw and his preliminary observations.

A recent television programme (Cutler, 1988) attempts to relate leukaemia clusters to proximity to industries discharging radioactivity, in particular the Capper Pass metal-refining factory near Hull. An investigation by the director of leukaemia research at Leeds University, however, cleared the factory of responsibility.

Most recently of all, a study at Edinburgh University suggests that children in remote areas such as Dounreay and Sellafield may be more vulnerable to certain viral infections which may be connected with leukaemia, because there is lower natural immunity in such isolated communities (Kinlen, 1988). Dr Kinlen compared the effects of population increases in the 1950s at the above two nuclear reprocessing sites with a similar population surge at Glenrothes in Fife, a rural area nowhere near a nuclear plant. There was a similar leukaemia cluster at Glenrothes between 1951 and 1987 - a fivefold excess over the statistically expected number. The town no longer has an excess of leukaemia cases, which Dr Kinlen suggests could have been a rare, abnormal response to a common infection.

Overall, the causes of the cluster effects are currently uncertain but if, pessimistically, the four Seascale leukaemia cases were all connected with 20 years of reprocessing Thermal Reactor fuels at Sellafield, the averaged death rate would still only be 0.02 per GW(E)yr of corresponding power.

Appendix 6.2

A COMPARISON OF HAZARDS FROM HLW WITH NATURAL RADIOACTIVE MATERIALS AND COAL-FIRED POWER WASTES

A.6.2.1 The Hazard Potential of Disposal in Saline Groundwater Relative to Natural Radioactivity

Here, the situation is assessed where the migration of nuclides eventually issues into the sea, not freshwater. A useful starting point to assess the resulting hazards is to compare them with those of natural radionuclides established in Appendix 1.2. The simplest nuclear fuel cycle, the Once-Through Cycle, yields the greatest quantities of long-lived nuclides, since the other cycles recover and burn up a proportion of the plutonium produced in previous cycles. For a plutonium-239 arising of 180 kgs per GW(E)yr from Table 2.1, the corresponding radioactivity is 410 TBqs (11,000

Ci): using ALI data from Hill (1981) with updating by ICRP (1986), the toxic potential for ingestion may be derived as 100 billion ALIs per GW(E)yr. With similar procedures for the other long-lived isotopes of Table 2.1, the following figures can be derived for the long-lived activity and ingestion toxicity of a Once-Through Cycle (per GW(E)yr) at (say) a century ex-reactor, when the majority of plutonium-241 has decayed to americium-241:

Table 6.3

Long-Lived Activity and Ingestion Toxicity per GW(E)yr of a Once-Through Cycle

Nuclide	**Pu239**	**Pu240**	**Pu242**	**Am241**	**Am243**	**Total**
Activity (TBqs)	400	600	2	6,000	20	7,000
Toxic Potential (billion ALIs)	100	200	0.5	700	3	1,000

Comparing the above totals with those for natural radioactivity in Appendix 1.2, it is found that the radioactivity of about 300 million/7,000 or 40,000 GW(E)yrs of spent fuel could be uniformly dispersed in the oceans before matching the natural radioactivity therein: such a fuel level is about 40 times that generated by world nuclear power to date. Similarly, using the figures for ALIs deduced in Section 5.4.2, about 300,000 billion/ 1,000 billion or 300 GW(E)yrs of spent fuel long-lived activity is equivalent to that of the natural toxic potential of radium in the oceans. The corresponding figure for the whole uranium-238 decay chain is 4,000 GW(E)yrs. The levels of both radioactivity and ALIs in spent fuel can obviously be substantially lessened if americium-241 can be prevented from entering the sea. The ALI level from neptunium-237, the decay product of americium-241, can be estimated as about ten times lower than that of americium-243 and so can reasonably be neglected.

Overall, therefore, a release into the sea of the total long-lived activity from nuclear fuel cycles, slow enough to avoid enhanced local seawater concentrations, would cause only a small fractional increase in the radioactivity and its associated hazard naturally present. A first target in disposing of HEW and spent fuel nuclear wastes is therefore to inhibit nuclide release for a few hundred years so that the relatively high level of radioactivity of the more important fission products does not reach the sea. An extension of this target to inhibit release of americium-241 for one or two thousand years could give a useful reduction in environmental impact. Such delays would almost certainly simultaneously provide a very slow release, thus precluding significant local impact, as can occur in (say) Sellafield pipeline discharges (Section 4.1.3).

The general philosophy above can be applied in part to release to fresh water, but the effects are considerably complicated by the arbitrary assumptions which have to be made of future drinking consumption.

A.6.2.2 **Comparative Assessment of Hazards from Nuclear and Coal-Fired Power Wastes**

In this section, we present assessments of the toxic effects of nuclear discharges relative to those causing the main hazards of coal-fired power, the stack emissions.

Table 6.2 gives the death rates of the general public (500-year to long-term) from coal-fired power: a middle-of-the-range figure may be taken as 100. Using the data of Camplin (1982) for sea discharges appropriate to Sellafield, updated as necessary by ICRP (1986), we may deduce a corresponding number of cancers per GW(E)yr caused by ingestion as 1 for plutonium isotopes and 9 for americium-241. This compares with about 50 for the important 30-year fission product caesium-137 at 10 years ex-reactor, falling by about a factor of ten each century. In other words, if all the long-lived activity in Once-Through Cycle spent fuel were discharged slowly from Sellafield, so that local hazards were insignificant, the ensuing death rate would only be about one-tenth the rate from coal-fired power.

An acceptable interpretation of the requirement that 'there should be no large-scale expansion of nuclear power until long-lived activity can be safely disposed' (RCEP, 1976), is that such activity from spent fuel is slowly released to sea, e.g. from Sellafield (and by inference from Camplin (1982) from many other places on the UK coast). Obviously, even safer disposal would occur if the majority of the activity from americium-241 could be decayed away before release; yet another useful factor of safety would be gained if plutonium was absent, as in HEW. Overall, therefore, it is sufficient for waste disposal designers to attempt to inhibit americium-241 from release to the sea, but the consequence of not fully achieving this aim is not serious. Even the slow release of a small fraction of the longer-lived fission products is not critical. Many of the disposal concepts for spent fuel and HEW can therefore be seen to be grossly over-designed. In fact, numerous assessments have shown that man-made barriers alone are sufficient to retain all nuclides for at least 1,000 years (Jones, 1987). Expensive research into a wide range of geological conditions is therefore not only difficult to assess, but unnecessary. The extra hazards in the preparation and transport of nuclear wastes over long distances through the public sector may in fact outweigh the postulated advantages of 'superior' geology (Black, 1980).

It may be concluded that, radiologically, it would be acceptable to dispose wastes in Dry Repositories draining to sea or in Dry Boxes of similar drainage, if the latter deteriorated only slowly after a century or two: remedial action could extend the integrity as desired. Retrievability is unnecessary. However, if there were appreciable fissile concentration in the waste, as in spent fuel, security against intrusion may indicate a preference towards disposal under the sea-bed (USL).

Chapter Seven

HEALTH EFFECTS FROM INFREQUENT AND MORE SERIOUS ACCIDENTS

7.1 GENERAL

Aspects of accidents with relatively slight consequences from power systems were outlined in the last chapter. In this chapter, more serious accidents are reviewed, including extensive damage to equipment and perhaps injury to large numbers of people, both operators and the general public. Because of the inherent infrequency of large-scale accidents, examples from worldwide occurrences have been recounted below. However, as a compromise between surveying a lengthy timescale to increase numbers of events and including only events relevant to modern practices, cases before the Second World War are excluded. This is also convenient because the use of nuclear power began post-war, since when it has generated roughly 100 GW(E)yrs of electricity in the UK, several hundred GW(E)yrs in the western world and around 1,000 GW(E)yrs worldwide. Deliberate sabotage will be omitted, partly because measures to counteract terrorism are not public knowledge. It suffices to say that as yet no power plant has been subject to a serious terrorist attack. Inherently, the thick concrete radiation shielding round the more radioactive areas of a nuclear plant provides considerable protection against explosives. Further, there are clearly more 'worthwhile', and easier, targets. Excellent discussions on terrorist attack and non-proliferation of atomic weapons are given by Cohen (1983) and Fremlin (1985), and on the general safety principles of nuclear operations by Franklin and Marsham in Foley (1978). In addition to reviewing past experience with nuclear plant, the nature of future nuclear accidents is deemed sufficiently important to merit a separate section for discussion.

7.2 A REVIEW OF SERIOUS NUCLEAR PLANT ACCIDENTS

7.2.1 Construction

No large-scale accidents have occurred during construction work on power or fuel cycle plants.

7.2.2 Fuel Preparation

No large-scale accidents or serious plant breakdowns have occurred during mining of uranium and its purification or conversion of uranium and plutonium compounds to nuclear fuel.

7.2.3 Power Operations

Below, the more serious nuclear power production events worldwide are reviewed in chronological order. The examples outlined typify the main bases of serious reactor accidents - fire, loss of cooling and an excursion into prompt criticality.

Perhaps the first serious accident to a reactor other than one used for research purposes was that in the UK at the Windscale No.1 Pile in 1957. Though the purpose of this reactor was to produce military grade plutonium and not to supply electricity to the Grid, the incident is instructive in its origin and effects on the environment. Briefly, the fuel was uranium metal clad in aluminium; the moderator was graphite, with cooling by air. Operator error at very low power caused too rapid a release of stored energy within the graphite, so that it overheated and caught fire in the air flow. There was no way to extinguish the fire other than to swamp it in a rapid flow of water. The stack filters removed most of the radioactivity reaching them: even so, it is estimated that about 260 persons could have contracted thyroid cancer from the intake of radionuclides of iodine, with perhaps 33 resulting deaths (Crick, 1983). Much fuel has since been removed and there has been a recent announcement (October, 1987) that the UKAEA will soon begin to take out the rest of the fuel and dismantle the reactor structure.

It is claimed, quite rightly, by the UKAEA that Windscale No.1 Pile was a military reactor of early design and that such an accident could not happen again on later military reactors at Calder Hall on the Sellafield site and Chapel Cross, near Dumfries which are cooled by carbon dioxide. This is circulated to yield up its heat and thus is recycled to raise steam to drive turbines, so supplying electricity to the National Grid. The civil Magnox reactors are scaled-up versions of Calder Hall. Advanced Gas-Cooled Reactors (AGRs) also have carbon dioxide cooling but operate at a higher temperature where the graphite moderator does not have the same stored energy properties.

The most serious US reactor accident occurred in 1979 at Three Mile Island, Pennsylvania. Here the main feedwater system malfunctioned, the auxiliary system switched on automatically as required, but valves had been left closed in both its lines. A pressure relief valve, which should have closed, did not do so, but indicated that it had. Water levels fell, exposing part of the core, which became sufficiently hot for zirconium cladding to react with the steam and generate hydrogen. The pressure relief valve was closed and the accumulation of hydrogen was gradually dissipated. The overall effect outside the reactor was calculated to be less than an even chance of one death from cancer. Pro-nuclear parties pointed to the unfortunate chain of events in spite of which there was little injury. Anti-nuclear groups claimed that it was fortunate that a much more serious accident had not materialised and that the cost of cleaning up the accident was astronomical.

The world's most serious nuclear accident occurred in late April 1986, at Chernobyl in the Ukraine. As at Three Mile Island, operator error was the primary cause. However, at Chernobyl the operators took premeditated action to sidestep the installed safety features in order to complete an experiment to programme. A failing in the design, which would not have been permissible under UK safety policies, was the positive temperature coefficient at low power, i.e. if the temperature rose, so did the power, the response being so rapid that the reactor went prompt critical (see Section A.1.4.2) before shutdown could be effected. An early estimate by Collier (1986) suggested that between 4,000 and 40,000 extra cancer deaths may eventually result over the next 40 years, about 1% of those to be expected through natural radiation. Gittus (1987), in a later more intensive investigation, gives a figure of 7,500 extra deaths for the European part of the USSR. On the other hand, an IAEA assessment (Novegno, 1987) suggests that only 2,900 extra cancers would result over the same region. A lurid account of events during the accident period is presented by the staff of the Observer newspaper in Hawkes (1986), but no immediate casualties other than to operators and firefighters were estimated. Greenpeace suggests an extra 100,000 deaths from Chernobyl, corresponding to 100 deaths per GW(E)yr worldwide, but the basis for the figure is not given. An interesting observation made in a 1987 World Health Organisation study is that the greatest potential health effects from Chernobyl would arise from caesium-137; effects from strontium-90 and plutonium would be insignificant.

In the assessment of casualties from reactor accidents, the handling of associated statistics is a matter of considerable debate. Here, a very simple approach will be adopted in setting out to obtain a 'feel' for the important implications. For a first estimate of casualty rates, based on UK experience only, the preponderant figure arises from the Windscale No.1 Pile accident. The only figure based on serious calculated techniques is that already quoted, i.e. a probable 33 deaths (Crick, 1983), so the Windscale accident corresponds to about 0.3 deaths per GW(E)yr in the UK. In the western world, the only incidents resulting in casualties have been well below those from the Windscale No.1 Pile. The death rate averaged over this region is therefore less than 0.1 per GW(E)yr. Reviewing worldwide experience, the casualties resulting from Chernobyl dwarf the total of all other accidents. Using the Gittus estimates above, with a slight increment to allow for deaths outside European USSR, an average of 8 deaths per GW(E)yr is obtained.

7.2.4 Reprocessing

There has been no large-scale accident during reprocessing or in waste disposal in the western world. In the USSR, there was a possible accident at Kyshtym in the Southern Urals. It is difficult to obtain a clear picture of the causes, since an official explanation has never been issued. A current hypothesis is that ammonium nitrate used in the extraction of plutonium was buried with the wastes, which were then dispersed when the ammonium nitrate dried out and exploded (chemically). As the above account of the Chernobyl incident demonstrates, it is perhaps unreasonable to include events in the USSR in estimates of average casualties from nuclear plant designed in the west, when the standards of Soviet design are different and often not at all clear.

7.2.5 Decommissioning

Little decommissioning of nuclear sites has yet been done, but the frequency and consequence of large accidents is unlikely to be important in the present comparisons.

7.3 THE CHANCE AND CONSEQUENCE OF FUTURE NUCLEAR PLANT ACCIDENTS IN THE UK

7.3.1 General

UK Government plans for the safety of the population after a serious nuclear incident include the prompt evacuation of persons within 2.4 kilometres of the scene. This compares with distances in US and Swedish arrangements of 16 and 40 to 80 kilometres respectively. At Chernobyl, the Soviet authorities moved people from within 30 kilometres of the Unit 4 reactor (Hawkes, 1986). Openshaw (1986), in a comprehensive review of reactor siting predating the Chernobyl incident, has pointed out that evacuation would be physically impossible for distances of tens of kilometres round UK reactors at Berkeley, Hartlepool and Oldbury, where the movement of about one million inhabitants could be involved.

7.3.2 Chernobyl-Type Accidents

The figure of 7,500 eventual excess deaths resulting from the Chernobyl accident, quoted in Section 7.2.3 from Gittus (1987), referred to the European part of the USSR. If such an accident took place in the UK to a reactor where movement of people within (say) 20 kilometres of the scene ('remote' siting) were feasible, the effects might be estimated, very roughly, to be similar to those experienced at Chernobyl. (There could, of course, be large variations, since many UK sites are on the coast and the direction of the wind could therefore be of critical importance.) Moreover, deaths from cancers induced by natural radiation in the UK over the next 40 years are estimated to be about 50,000 - roughly 1% of all cancer deaths (and 0.3% of deaths from all causes). The consequences of 'Chernobyl UK' would therefore be difficult to distinguish from natural effects. To put the above consequence of Chernobyl in another perspective, we can deduce from the 50 deaths per GW(E)yr at current UK coal-fired power levels of about 40 GW(E), that there are some 2,000 deaths annually from this source, i.e. a few years with the present power 'mix' dominated by coal is equivalent to 'a Chernobyl'. Turning the issue the other way round, the steady death toll from coal-fired power might be regarded as an insurance premium aginst low-frequency nuclear disaster. The figures presented in this book would suggest that such a premium was much too high.

Our crude 'historical' guide from casualties per GW(E)yr to date does not, of course, take account of lessons learned in the decades of experience of nuclear power, which will be applied to improve safety in the future. Many detailed articles have been published assessing probabilities and counter-measures related to accidents. We have only applied a 'broad brush' approach here, but the question of whether to include Chernobyl in evaluating the future safety of western world reactors can be seen from the above discussion to be important. It is certainly true that the Chernobyl-type design would never have been contemplated in the UK. For example, an attempt was made in the 1960s in the UK to design an SGHWR-type reactor which could operate on natural uranium; an important reason why the study was discontinued was the inability to find an arrangement with a negative temperature coefficient of reactivity. The use of enriched uranium, as in the Winfrith SGHWR, which has operated very satisfactorily for 20 years, is in fact one of the measures taken to correct the basic operation of Chernobyl-type reactors.

Gittus (1987) summarises the position as below:

> UK safety rules first and foremost aim at the building of reactors that have intrinsic characteristics that provide inherent protection. Secondly, these natural defences are supplemented by engineered features to prevent, limit, terminate and mitigate any faults. Thirdly, the systems design must be tolerant to operator action - if the operator makes a mistake, the reactor shuts down. Fourthly, UK operators are highly educated and well trained, not just for routine operations, but for unusual and accident situations and, fifthly, the entire system is overseen by an independent Nuclear Inspectorate that can at any time, without hindrance or challenge, close down any licensed reactor ... The reasons for the accident at Chernobyl are now clear. It occurred as a result of three main design drawbacks of the reactor:
>
> 1. it had a positive void coefficient (the reactivity increased if the gas volume in the reactor increased) and, below 20% power, a positive power coefficient, which made it intrinsically unstable at low power;
>
> 2. the shutdown system was in the event too slow in its operation;
>
> 3. there were no physical controls to prevent the staff from operating the reactor in its unstable regime or with safeguard systems seriously disabled or degraded.
>
> (Gittus, 1987)

7.3.3 **Three Mile Island-Type Accidents**

The most likely serious accident relevant to future UK power, but still of very low probability (the Three Mile Island occurrence is the only one so far recorded), is where there is an appreciable release of activity into the pressure vessel of a PWR. Though there is no detriment to the public, the reactor may then be put out of action permanently and decontamination deferred for many years. The situation has some similarity, but to a more severe degree, to a reactor routinely decommissioned to Stage 3; in other words, there is negligible hazard outside the reactor shell, but further operations are rendered much easier by delaying them for a long period, such as several decades or even a century.

7.3.4 'Short-Term Accidents'

In this group, we gather together the small chemical explosions which result in an aerial spread of contamination within a limited section of a fuel plant or reactor. It is then practicable to clean up affected areas and restart the plant. If short-lived isotopes are dispersed, as was ruthenium-106 in B204 at Sellafield in September 1973 (see Appendix 6.1), a decision on whether to decontaminate and restart the plant will depend on the availability of alternative plant. For example, a delay of 10 years would reduce contamination by ruthenium by a factor of a thousand. On the other hand, the value of restarting the affected unit quickly may justify expensive decontamination. In any event, the incident is likely to have negligible effect on the general public.

7.3.5 Other Hypothetical Events

The accidents experienced with nuclear plant so far have arisen from errors in their design or operation or both. To date, low-probability events have not been the basic cause. Nevertheless, their effects can be sufficiently serious that they must be considered during the design stage.

Among the most unlikely natural hazards are new volcanic eruptions and impact with a large meteorite. These are such momentous events that there is little that engineering safeguards can achieve against a close encounter. The geological history of the UK, however, shows that such events have not occurred for millions of years and the likelihood of a given plant sustaining a close impact, even during (say) several centuries or more, while nuclear waste was potentially hazardous, can reasonably be discounted. Furthermore, the nuclear factor would add only a small extra hazard relative to the primary event.

More distant effects from volcanoes and meteorites can be considered as a class of earthquake. The latter, of course, occurs much more frequently than volcanic eruptions and large meteorite impacts. Even in the UK, earthquakes of varying severity have been experienced since records began, about 200 years ago. Considerable attention is therefore paid in design to counter earthquakes to standards laid down by NII. Much useful information on the strength of structures during earth movements can be derived from studies of other regions of the world more prone to this type of event, e.g. Japan (Burton, 1981).

Regardless of the cause of a nuclear accident, the inherent property of continuing heat output from nuclide decay can create severe secondary effects. If the heat is not removed fast enough, temperatures can arise causing structural materials to melt and sections of plant can, in principle, coalesce in a 'meltdown'. If a critical mass is formed, then explosions will disperse the material, as happened at Chernobyl. These explosions will not be like atomic bomb reactions, since the dispersal takes place, as at Chernobyl, before sufficient energy is generated by fission (Section A.1.4.2). All of the noble gas and a large part of the volatile fission products were released at Chernobyl. The environmental effect of a meltdown would not be much worse, since the dose to large populations was dependent on the volatile fission product, caesium-137. Continued melting of structural material will eventually form a surface large enough to dissipate heat without further temperature rises and melting. Meanwhile, chemical reactions, say with groundwater, could create Chernobyl-like effects and associated dispersal of the melt. Arrangements for dropping neutron pickup materials, such as compounds of boron, into a reactor under threat of an uncontrollable excursion, are among precautions which may be included in design.

7.3.6 <u>Remote Siting</u>

Only the major reactor accidents are of sufficient importance to influence the choice of plant siting. There is clearly some benefit in cost and operator exposure in being able, after routine decommissioning to Stage 3, or after serious contamination within a reactor shell, to seal up the active region for very long periods if not indefinitely. This may induce planners to prefer remote siting to cope with the possibility of tens of 'dead' nuclear reactors not far from centres of population in the future. A spin-off from this policy would be to reassure the public, not only against the more serious local consequences of a Chernobyl-type incident, but also, the sensationalising in the media which no doubt will be generated in the eventual 'burning' of plutonium as a source of power in either Thermal or Fast Reactors.

In reality, it is difficult to plan emergency evacuation effectively. In the first place, the admission of the possibility of such an emergency reduces public confidence. In the second place, as experienced at Three Mile Island, large groups of population will instinctively decide for themselves to leave the area, probably causing more havoc and casualties than the nuclear incident itself. On the other hand, the public can easily 'see' the extra degree of safety afforded by remote siting, whereas engineering safeguards are less easily understood and can be swept aside with such phrases as 'what can go wrong, will go wrong'. There is therefore a considerable inducement towards remote siting of future nuclear plants.

7.4 COAL-FIRED POWER

7.4.1 Construction

As with nuclear plants, no large-scale accidents have occurred during construction of many mines and coal-fired plants in the UK since the Second World War.

7.4.2 Fuel Preparation

Mining is well known as a hazardous occupation, though in fact it is no more so than in heavy engineering industry generally. Two large-scale mining accidents resulted in 104 deaths at Whitehaven in 1947 and 83 deaths at Easington in 1951 (HSC, 1978).

Important hazards to the public from coal mining come from the heaps of spoil and the lagoons resulting from excavation and coal cleaning. In the terrible accident at Aberfan near Merthyr Tydfil in October 1966, an avalanche of coal slurry from Tip No.7 engulfed Pantglas school, killing 116 pupils, 23 teachers and 23 other adults. Since then, new regulations for coal spoil heaps have reduced their steepness: no further accidents of this type and on this scale have occurred. Assuming about 1,000 GW(E)yrs of coal-fired power since the Second World War in the UK, the probable operator death rates, based on the Whitehaven and Easington accidents, would be 0.2 per GW(E)yr; the corresponding figure for the public based on Aberfan would be about the same.

7.4.3 **Power Operation**

In Chapter Five, the health effects of fossil-fuel combustion products discharged to the atmosphere were discussed. Normally, discharges from household chimneys and power station stacks rise high enough to give dispersion. In unusual weather conditions, the discharges sink rather than rise, so that high concentrations of particulates and combustion gases exist at ground level.

In December 1952, the latter type of weather lasted in London for several days: the resulting 'smog' was estimated to have caused 4,000 excess deaths in one week. Though domestic fires contributed to casualties, and these would have tended to occur with older people with respiratory problems, many more were likely to die prematurely through bronchial effects in subsequent years. Other London episodes in the period 1948 to 1962, though smaller than that of December 1952, each resulted in hundreds of excess deaths. New York, too, had three similar disasters in 1953, 1963 and 1966, resulting in 360, 500 and 160 estimated excess deaths. Averaging the UK figures, the 'smog' incidents would be equivalent to about 5 deaths per GW(E)yr. However, it may not seem reasonable to use this figure for future predictions, since the regulations on smoke abatement have done much to reduce the frequency and severity of smogs in the UK. Since the above figure is ten times lower than the 'routine' level of 50 deaths per GW(E)yr of Table 6.2 for coal-fired power, it would seem sensible to regard any residual smog casualties as part of 'routine' effects.

7.4.4 Decommissioning

No large-scale accidents have occurred in demolishing power statons fired by coal or in closing collieries.

7.5 FUSION AND RENEWABLE POWER SYSTEMS

Hazards to the public from a fusion power accident should not be severe - the reactor only contains a minute amount of fuel and the reaction cannot 'run away'. Moreover, the absence of fission products and heavy isotopes would make the consequences much less serious than for fission power. There would also be less decay heat after shutdown and so slower and less warming up of structural components. There is no possibility of subsequent criticality and so a 'meltdown' is precluded.

Large-scale hazards to the public from wind, wave and geothermal power are also difficult to postulate. All could be sited well away from public access or at least enclosed. Perhaps the most likely hazard is from collision of ships with wave machines, but there may be few deaths in such an eventuality. On the other hand, exceptional weather conditions in the North Sea capsized the oil rig Alexander Kjelland in March 1980, with the loss of 100 lives. Tidal power is more akin to the nuclear system in that the probability of a sudden major rupture of a barrage is very small, but the consequences are very serious (Henderson, 1987). The failure of the Gujarat Dam in India in 1979, for example, killed about 15,000 people. This dam was on a river, and the population living in jeopardy from a tidal dam would possibly be smaller, but the chance of structural failure would seem the same. It follows that, if Chernobyl, with its 'UK-unacceptable' design and operation, is taken as a reason for abandoning nuclear power, it might be argued that tidal power should also be excluded from future UK power mixes because of the failure at Gujarat.

7.6 SUMMARY

It is not the intention of this book to assess the technical aspects of large-scale reactor accidents: the factors involved are too complex to recount here. Several excellent texts on the more serious accidents have been written (Windscale, Three Mile Island and Chernobyl). The accidents themselves have to a considerable extent been practical demonstrations that a nuclear power station cannot explode in the manner of an atomic bomb. The problems which can be caused by reactor accidents are the secondary chemical reactions - burning of graphite and uranium metal (Windscale), formation of hydrogen bubbles (Three Mile Island) and prompt criticality explosions followed by high-temperature reactions of zirconium and graphite with steam or air (Chernobyl). Similar such chemical effects would probably follow low-frequency natural events such as earthquakes, including approaches to meltdowns. Anti-nuclear groups link all of the accidents together under the word 'nuclear' and prophesy further disasters. Pro-nuclear organisations point out that shortcomings of design and of operational procedures have been eliminated in existing or new designs.

In Section 7.3, and in a later section of this book, possibilities of reducing the consequences of serious reactor accidents by careful siting are put forward. Meanwhile, it is worthwhile noting, by comparing figures in Tables 6.2 and 7.1, how small the averaged death rate of most serious accidents is, relative to the 'routine' deaths from atmospheric pollution evaluated in the last chapter. Though our averaging method for accidents is very crude, it is likely, if anything, to exaggerate death rates. Based on UK accidents only, it can be seen that, even if the Windscale fire is included, large-scale average death rates are in single figures for nuclear power systems and a hundred times smaller than routine casualties from coal-fired power. When Chernobyl is taken into account in a worldwide survey, the rate only increases to a similar level to deaths from radon in UK houses (Section 5.4.3) and is well below the death rate from coal-fired power station discharges to the atmosphere. The 'far field' effect of a 'UK Chernobyl' would, in fact, have a similar long-term hazard to current levels of coal-fired power with the difference that the toxic discharges from the latter are not easily detectable. Overall, there would appear to be less reason to take Chernobyl into account in predicting future casualties from nuclear power than the inclusion of smog casualties of the 1950s into future effects of coal-fired power.

In nuclear fuel cycle accidents, the effects are even smaller, since the energy potentially available for release is much less than in reactors. In spite of a long campaign by the media and anti-nuclear organisations to suggest that Sellafield operations are dangerous, the effects of the latter are quite negligible.

Fusion systems are inherently safer than fission systems. Most renewable systems, needing little or no fuel and creating only small amounts of waste, appear to have little potential for large-scale accidents. However, a sudden rupture of a tidal barrage, either by accident or by sabotage, could have serious consequences.

Table 7.1

Summary of Serious Accident Death Rates Averaged per GW(E)yr for Various Power Systems (a)

System/Phase	Nuclear	Coal	Fusion	Wave	Wind
Construction	vs	vs	vs	vs	vs
Fuel Preparation	vs	0.2(b)	vs	-	-
Power Operation	(c)	vs	vs	vs	vs
Reprocessing	vs	-	-	-	-
Waste Disposal	vs	0.2(h)	vs	-	-
Decommissioning	vs	vs	vs	vs	vs

Notes: vs denotes very small, i.e. less than 0.01.

(a) All estimates refer to casualties to the public, except (b).

(b) Serious mine accidents averaged over total UK post-war power.

(c) Figures here are 0, 0.4, very small and 8 for conditions of notes (d) to (g) respectively.

(d) Averaged over UK nuclear power, excluding the Windscale disaster.

(e) Averaged over UK nuclear power, including the Windscale disaster.

(f) Averaged over world nuclear power, excluding Chernobyl.

(g) Averaged over world nuclear power, including Chernobyl.

(h) Aberfan deaths averaged over UK post-war coal-fired power.

Chapter Eight

SYSTEM COMBINATIONS FOR FUTURE UK POWER SUPPLY

8.1 GENERAL

Since the Second World War, the annual electricity consumption in the UK has increased roughly fourfold, with the highest increase occurring in the south. Superimposed on the average rate of increase are considerable short-term fluctuations. More electrical heating is required in winter than in summer and the demands of industry in daytime are higher than at night. Additionally, there are very short surges in demand, a well-known example resulting from the switching on of numerous kettles and electric rings during the commercial breaks in a James Bond film. Such demand characteristics reflect back on the choice of power 'mix', since some types of power station are more economic to run steadily on 'base load', whereas others are more suited to cope with short surges in demand.

Historically, it has always been difficult to forecast year-by-year changes in power requirements and many projections have eventually been found to be quite wide of the mark. For example, the Arab-Israeli War of 1973 resulted in a fivefold increase in the price of oil; this was, of course, virtually impossible to predict, yet had an important effect on power policies. A similar uncertainty occurs in the cost of uranium for nuclear fuel. With the present apparent abundance of uranium, the ore price is relatively low, but it is not improbable that a sudden world impulse to stockpile it for the future could multiply prices severalfold. In the next section, we shall be discussing the factors influencing the forecasting of the size and pattern of future overall electricity supply. In order to set the scene for discussions on which combinations of systems could best satisfy these forecasts, the plant costs for construction, operation and decommissioning of power systems are discussed in outline in Section 8.3. In Section 8.4, the detrimental effects on the environment through loss of health, damage to buildings, etc., are evaluated in equivalent financial terms. Salient points arising from a comparison of total system characteristics to satisfy the projected demand are then assembled in Section 8.5, as a basis for a discussion on the projected power mix. Finally, key points from the foregoing sections are summarised in Section 8.6.

8.2 THE PREDICTION OF FUTURE DEMAND PATTERNS

Most of the current UK power supply is derived from coal (82%), with the preponderance of the rest being nuclear (15%). Such a dominance by coal has induced doubts on the reliability of supply, mainly because of the recurrence of strikes in the mining industry every few years. Both the increase and change of pattern in demand since the Second World War have been dependent on a number of factors, such as changes in industrial processes, stricter smoke emission regulations and the replacement of fossil fuel burning for domestic purposes by labour-saving electric heaters. There are several other important factors influencing the structure and totality of future power demand.

(a) A high unit cost of electricity can influence customers towards greater conservation, as in better thermal insulation of property, more efficient equipment in the kitchen, etc. Department of Energy predictions suggest 5-20% of the UK consumption of energy (not necessarily as power) might be saved in this way (Section 1.4.4).

(b) If electricity installations have already been built to provide sufficient supply in peak periods, increments in off-peak sales effectively have a production cost corresponding mainly to the increment in fuel usage, since capital costs have to be paid whether plants are being operated or not. Particularly advantageous off-peak prices can therefore be offered, inducing increased electricity consumption. An interesting has occurred in France, where an excess of nuclear capacity has recently developed. Since, as will be described later, the operating side of nuclear power is relatively inexpensive, there has been a campaign by the French electricity supply industry to induce a greater off-peak (and overall) consumption through price reductions. Partly as a consequence, most new French houses are all-electric, a result which will thus be reflected in increased future off-peak demand.

Another effect of the preponderance of French electricity stemming from nuclear power is that, apart from Norwegian hydropower, it is the cheapest in Europe. French industry therefore stands at an appreciable advantage with respect to the rest of Europe, in particular the UK. Lower UK production, through being less competitive, then reflects back on lower requirements for power.

(c) On the other hand, in order to take part of the load during brief periods at peak demand, it may be advantageous to use plant with low capital costs. It has been customary in recent years to use gas-fired plants in this way, even though these are three times as expensive to run on base load as nuclear plants (Blowers, 1987). Moreover, gas turbines and pumped storage systems can be brought to full power in seconds whereas large nuclear and coal-fired plants take hours. The CEGB does in fact have a ranking order, running nuclear plant on base load, the newer and more efficient coal-fired stations next, and so on, until the gas turbines are brought in for 'peak-lopping'.

(d) Some changes in demand structure may be worthwhile as older plants are considered for replacement. Here, the reliability and costs of continuing to run such plants must be weighed against the costs of both building and running new plants.

(e) The privatisation of electricity supply, under implementation by the present UK Government, could, in the long term, change present policies from siting coal-fired power stations near to large coalfields to proximity to deep-water ports, where the economics of imported coal could be attractive. Since the average cost of coal mined in Britain is currently about twice the world spot price, use of imported coal could well swing the balance against nuclear, at the same time increasing overall power usage. On the other hand, since significant imports of coal would be one of the few developments which could unite all the miners' unions into national strike action, this is a situation which any UK Government would be anxious to avoid. Further, imports of coal on a large scale could have a serious effect on the balance of trade, which is already badly in deficit. Moreover, it seems probable that alternatives to coal would be more likely to be preferred to increase power production in the south and thus decrease transmission costs.

Forecasting electricity demand is a notoriously inaccurate business: because of the long construction times of power plant, it is not easy to change tack and the predictions need to extend for several decades. The past efforts of the CEGB and other Establishment bodies have been so far from the eventual realisation that they have been the subject of much ridicule by the media and environmental groups. A typical example of such predictions (UKAEA, 1974) suggests a level of nuclear power alone of over 40 GW(E) by the end of 1990, a factor of about three over the current likely outcome. Most of such forecasts, up to the 1970s, were based on the assumption that there was a close association between electricity demand and Gross Domestic Product (GDP), which is a reflection of the standard of living. Such a correlation was derived from data of power usage against GDP for many countries around the world. However, in the UK there has been a fluctuating rate of growth of GDP and also its composition, as the economy moves from traditional manufacturing, involving heavy electricity consumption, to less energy-intensive, service-type industries. There is also a higher level of finished imports, which effectively transfers domestic energy consumption overseas.

At the Sizewell Enquiry, the Department of Energy presented eight scenarios of total power demand and system mixes up to the year 2010 (DoEn, 1982); case D was the one most favoured by the Department and the CEGB. In this, demand was expected to increase slowly, by about 25% from 1982 to 2010, *inter alia*, because of a low annual increase of GDP (0.5%), low industrial growth and high energy prices. Objectors at the Enquiry queried even this modest increase, partly because of the data used in the forecast and partly on the grounds that conservation measures could make large reductions in demand.

Overall, the above differences in the eight demand scenarios are not great, mainly affecting the date of the construction of Sizewell, not the long-term structure of power system types. Consequently, in our analyses in the remainder of this book, it seems reasonable to assume that demand may not grow (or decrease) drastically in the next few decades, in line with the above scenario D. However, there is a small, but not negligible, chance that, in the long term, large changes could occur, so that it will be important to examine the flexibility of proposed power mixes to such gross effects.

8.3 ASSESSMENT OF PLANT COSTS

8.3.1 General

The descriptions in earlier chapters of the various power systems for the UK have given an outline both of the technical features of their operation and of the impact of their discharges on the environment, in particular, the damage to the health of operators and the public at large. In order to determine preferences between systems, a method is required for comparing the above environmental effects with various monetary flows involved in the power plants themselves. Generally, this is done by assigning a monetary value to the various cost characteristics and 'discounting' each value back to a reference point in time.

In this section, the basic principles of 'discounting' are first described: next, we discuss the manner in which this is used to assess costs arising in the three phases of plant life - construction, operation and decommissioning. Figures are then presented for plants using the discounting procedure set out by the CEGB at the Sizewell Enquiry in their comparison of nuclear and coal-fired power costs, followed by reviews of the main points of objectors thereat and then the conclusions of the Enquiry Inspector. The final part of this section gives some indicative costs for fusion and renewable systems.

8.3.2 Costing Procedures

A plant required to provide power to a given programme will naturally receive an income from electricity sales which is unaffected by the type of its energy source. An economic comparison between systems can therefore be obtained merely by consideration of the costs of the systems, regardless of the unit price of electricity. Such costs include,

(a) investment outflow, which must be paid out as each construction stage is completed,

(b) expenditure due to operations through the useful life of the plant, and

(c) further costs which are incurred as the plant is decommissioned (less any receipts for saleable scrap).

In order to connect such expenditures arising at different times, it is common practice in industry to 'correct' a future cost, using an agreed rate of interest, back to a reference date, a procedure known as Discounted Cash Flow (DCF) analysis. Since this date is often the present, the 'corrected' figure is known as the 'Present Value' (PV). The interest rate used is then referred to as the discount rate. For example, £30 invested at an interest rate of 5% in 1990 would grow to £100 by the year 2005. Alternatively, it can be said that an expenditure of £100 in 2005 can be 'discounted' back to give a PV of £30 in 1990. Some PVs of £100 discounted back to 1990 at an interest rate of 5% (with the figures for 10% in brackets) from various points in time are listed below:

Year	**1990**	**1995**	**2005**	**2090**
PV (£)	100(100)	78(62)	30(9)	0.8(0.007)

Clearly, costs incurred early in plant life appear relatively more important than similar sums incurred later on. The effect is accentuated when the higher interest rate of 10% is used.

The consequences of these effects can be exemplified by a broad comparison of coal and nuclear systems of similar power output. It is generally accepted that the coal plant will cost less to build but will be more expensive to run than the nuclear plant (see Table 8.1 in the next section). The running costs, however, are incurred at a later time and continue over several decades. The coal plant 'gains', therefore, in a PV exercise because the higher running costs, incurred after the construction costs, are reduced in importance by discounting. Such effects are even more marked, of course, as the interest rate used increases. Briefly, coal benefits from the reversed telescope effect of discounting.

Longer construction times before a given startup date imply that the PV of expenditure incurred in construction is increased relative to later costs. Historically, nuclear plants have, on average, a poor record in this respect. Some plants have been built to programme, e.g. the Winfrith SGHWR was completed in four years in the late 1960s and an appreciable proportion of the allotted funds handed back to the Treasury. Mostly, however, nuclear plants have overrun programmes by several years, perhaps the worst instance being the Dungeness B reactor with a construction time of 17 years (Ince, 1984). Coal plants in general have been quicker than nuclear to build, though there are instances, such as at Ince, near Chester, and at the Isle of Grain, where construction timescales proved to be substantially greater than first projected.

Long timescales, such as a century, can be seen in the above table to discount costs downwards by a very large factor. Such severe discounting is probably unimportant when the scrap value of the plant is not much less than the cost of the residual operations. However, in the nuclear field, there is relatively little worthwhile scrap in the more radioactive part of plants, and the disposal of HLW and the Stage 3 decommissioning of reactors could occur as long as a century after plant shutdown. The problem is complicated in several ways. In the case of HLW, burial becomes less difficult with time as important nuclides decay away - caesium-137 and strontium-90 both are reduced by a factor of about 2 every 30 years. Not only do hazards to operators (or costs of equipment) reduce correspondingly with time but, as explained in Appendix 3.1, the amount of excavation required for disposal decreases as the decay heat output falls. There is therefore a real reduction in cost in delaying disposal, even after allowing for the extra costs of longer storage. Delays of up to a century in carrying out Stage 3 decommissioning of reactors provide similar reductions in the hazard potential - cobalt-60, with a half-life of just over five years, can effectively be completely decayed away. Overall, it would seem unrealistic to discount such long-term operations so drastically: a lower or even zero rate of interest would perhaps be more appropriate. On the other hand, this would be inconsistent with respect to comparing running costs during power production, where the advantage of nuclear over coal would increase markedly at lower interest rates.

In the next section, the CEGB's cost comparisons produced at the Sizewell Enquiry are presented. A point of interest concerning the 5% discount rate, recommended by the UK Government and applied by the CEGB at Sizewell, is that it may be below the interest which a private investor could receive from a building society, even when the latter has been reduced to allow for inflation. What therefore may be a sound criterion for Government decisions may be of less concern to a private investor. There are consequent problems in this respect for nuclear power in attracting investors if the Government's plans for privatisation of the power industry go ahead (Pryke, 1987).

8.3.3 The CEGB Comparison of Nuclear and Coal-fired Power at Sizewell

The CEGB's strategy of 'ranking' plants has been mentioned in Section 8.2. The issue at Sizewell, however, concerned only the type of plants to be preferred for future base load operation. For such plants, the CEGB prepared a comparison of costs of alternative future power production methods in a published booklet (CEGB, 1985): this was then used at Sizewell as the cost basis for the CEGB case. The three systems assessed were: a PWR nuclear reactor, an AGR nuclear reactor and a modern design of coal-fired power station, excluding FGD plant and fuelled with UK coal. Fusion and renewable systems were not included in this comparison because 'they are only in their formative stages and no basis exists for providing any meaningful cost comparisons'. Illustration VII from the booklet is reproduced below in Table 8.1; the discounting is at 5% to a reference date of March 1984.

Table 8.1

Generation Costs (p/kwhr) over the Whole Lifetime of Future Power Stations

	Sizewell B PWR	**AGR Station**	**Coal-Fired Station**
Capital Charges	1.99	2.46	1.20
Fuel Costs	0.59	0.87	2.81
Other Operating Costs	0.36	0.34	0.28
Total Generating Costs	2.94	3.67	4.29

The figures show that future coal power stations will generate electricity at almost 50% greater cost than PWRs. Further, the figures for the AGR and PWR in the above table show that about 70% of total costs are derived from capital charges. The position is roughly reversed for coal-fired stations, where a similar percentage arises from running costs.

In a variant of the above analysis, the three types of plant were tested as to whether they would be economic to commission, relative to continuing to run average existing base load coal-fired plants (Illustration VIII of CEGB, 1985). Whereas an AGR was found to be just economic, the PWR was found to be clearly economic, but the coal-fired station was found to be clearly uneconomic.

Before we examine nuclear costs in a little more detail, it is necessary to define the scope of the term 'fuel costs'. For fossil fuels, the definition relates straightforwardly to the raw material coal input. For nuclear fuel, however, it includes the ore concentrate as bought, plus the fuel cycle costs of chemical purification, enrichment of fissionable isotopes, fuel manufacture, waste disposal and (except for a Once-Through Cycle) reprocessing. According to BNFL, the latter accounts for only 20% of the total fuel cycle costs; however, French reprocessing costs have been quoted as four times those given at the Enquiry. Such an increase applied to the Enquiry figures would reduce the advantage of nuclear over coal substantially and provide more pressure for reprocessing to discontinue. On the other hand, the CEGB claims in its Enquiry evidence to have found that the conclusions of its analysis could stand up to a variety of such possible variations. The costs of the alternatives of the burial of spent fuel or its long-term storage are discussed in Section 8.5.2; these indicate that some saving of fuel cycle costs would be made, but the discounting of the costs of reprocessing carried out in the distant future appears as dubious as a similar procedure for decommissioning, mentioned above. Moreover, even without discounting, reprocessing operational costs can look small when compared with the cost of the equivalent electricity produced. For example, a disposal cost of HEW glass blocks of £0.5m per cubic metre, arising at 5 cubic metres per GW(E)yr (Table 3.1) is only equivalent to 0.03p/kwhr of corresponding power.

Overall, the conclusion by the CEGB that Thermal Reactor power systems are cheaper than coal-fired power systems is in line with those of many similar analyses carried out in European countries. Corresponding comparisons in the US yield a much closer balance, mainly because of the much lower cost of coal.

The CEGB excluded Fast Reactor costs from its presentation because long-term predictions were not considered. It is interesting, however, to note that current EEC studies suggest that commercial Fast Reactor system costs could be similar to those of established PWR systems at present uranium prices.

Before leaving the subject of comparative power costs, it is interesting to recall the commercial reaction known as the 'Sailing Ship' syndrome. When steam began to replace sail as the propulsion for merchant ships, there was a marked improvement in the performance of sailing ships. Such a reaction follows the introduction of many new commercial processes, since the old process will charge 'what the market will bear'. In the power context, coal costs will be cut to improve competitiveness with nuclear power, whereas the latter will charge as high a price as will keep it just economic with respect to coal. Benefits in fact will be channelled into the nuclear industry rather than to the consumer. The full benefits of nuclear power may come only when there is serious competition, say, from cheap imported coal or a process not yet conceived.

8.3.4 The Views of Objectors at Sizewell

The main points querying the CEGB presentation are set out below.

(a) AGRs might be a more prudent nuclear 'buy', since PWRs are still an unknown quantity (in the UK). In particular, the SSEB claims that AGRs could provide electricity about 4% cheaper than PWRs.

(b) Construction times for AGRs had generally been two to three times those forecast, so that the construction of PWRs would probably take as long as the US average of 11.5 years, rather than the forecast 7.5 years. (Because of the high capital component of discounted nuclear costs, they are sensitive to construction overruns.)

(c) CHP could be very advantageous in some parts of the UK, reducing significantly the power requirements from other sources. (The CEGB counter-argument is that this would affect mostly smaller stations in the 200 MW range, below the level they use currently.)

(d) The coal prices and money exchange rates assumed by CEGB could be sufficiently inaccurate as to cut substantially the advantage claimed for the PWR over alternative plants.

(e) The availability of a PWR to produce power to demand and the lifetime of the reactor will be less than assumed by CEGB.

8.3.5 **The Conclusions of the Enquiry Inspector at Sizewell**

With a few preconditions on matters such as operator training and 'tolerable levels' of risk to operators and the public, the Inspector, Sir Frank Layfield, accepted the CEGB arguments on the safety of Sizewell B. On other aspects, he felt the CEGB had 'significantly overestimated the likely future price of heavy fuel oil and coal'. On construction times, he felt that, in spite of the relatively poor past reactor construction record in the UK and overseas, the CEGB would have learned considerably from its experience. Though the cost-saving case did not seem as robust as put forward, his judgement was that

(a) there was a chance of about one in four that Sizewell B (PWR) would not be cost saving; the probability that an AGR would have lower costs than a PWR was about one in five,

(b) the chance that a coal-fired plant (with FGD equipment) would have lower costs than a PWR was about one in forty,

(c) the export market for PWRs was likely to be modest and not approach the forecasts of CEGB,

(d) there was no reason why Sizewell B and its successors should not proceed at the same time as increased conservation measures, and

(e) there was only a low probability that, if Sizewell B were to be started in 1987, it would be found to have been constructed much ahead of the need for new capacity.

Although the Enquiry did not address the prospects for future PWRs in the UK, the normal 'learning curve' from the first few reactors should suggest that a further cost advantage over alternatives will accrue with experience. The flexibility in the siting of nuclear plants compared with those based on coal should also be as good as that of Sizewell, since there will still be an imbalance of supply requiring transmission to the south, where there is relatively little coal production. (Currently, about 9 GW(E) is transmitted from the Midlands to the south, not far short of the practical maximum of 10 GW(E).)

8.3.6 Plant Costs for Fusion and Renewable Power

None of the major objectors at Sizewell chose to make out a case for renewables as a serious alternative to nuclear power (see Ince (1984), McKerron (1984) and Pryke (1987)). Predicted costs for these processes and fusion must therefore be deduced from other sources.

Recent OECD/NEA studies predict that generating costs of fusion reactors will be about 50% greater than for fission reactors. However, the dominant function, the capital cost, is strongly dependent on scale, so that units of 1.2 GW(E) or larger would be needed to achieve reasonably low generation costs.

A review of wind power (Milborrow, 1985) suggests, for average onshore sites, costs in the range 3.2p to 3.5p/kwhr, i.e. on the CEGB Sizewell figures of Section 8.3.3, wind power would be cheaper than coal but dearer than PWR power. However, the number of such sites is limited and costs increase substantially as offshore 'farms' are considered. The recent Severn Barrage study by McAlpine concluded that 5% of UK power could be supplied at a similar generating cost to Sizewell B. On the other hand, a 1982 Department of Energy study concluded that the best form of wave power is unlikely to produce electricity at less than 5p/kwhr. The Camborne School of Mines investigations of 'hot rocks' suggest a figure of 4.2p/kwhr for geothermal power in favourable areas if heat extraction at 6-kilometre depths were to prove feasible (Batchelor, 1985).

The above extra cost of renewable systems over nuclear power might seem small, but an increment of 1p/kwhr on 10 GW(E) output could entail an extra overall annual cost of about £1,000 million!

8.4 VALUATION OF ENVIRONMENTAL IMPACTS

Kindly Leave This Planet - As You Would Have Found It! This is a familiar saying among waste disposal workers and assessors of environmental impacts. It is a plea which is often cited or ignored according to what is required to be proved at the time. In this section we try to set values to present and future environmental consequences of competing power systems, so that their importance relative to plant costs can be set in perspective. First are reviewed some statistics on the risks of various human activities and estimates of the corresponding effects on life expectancy. A discussion of modern attempts to value the loss in the quality of life from diseases and disabilities, whether or not resulting in death, is followed by simple valuations from actuarial data. These discussions allow figures to be derived for the impact on human health of the various forms of power system. Impacts on other animal life, flora and artefacts, are then also discussed and quantified.

O'Brien (1986) quotes some risks estimated to increase the chance of death in the US in any year by one part in a million. Smoking 1.4 cigarettes is one such activity with the risk of incurring cancer or heart disease at the above rate. This basic figure may be used to transpose data from this reference for some other common activities into the same units as the risks from various forms of radiation given in Table 5.1, i.e. the effect of smoking one cigarette.

Table 8.2

The Relation Between Various Activities And The Risk Of Death

Activity Creating the Equivalent Risk to Smoking One Cigarette	Cause of Death
Spending 40 minutes in a coal mine	Lung disease
Travelling 7 miles by bicycle	Accident
0.7 chest X-rays (average)	Cancer
Living 6 weeks with a cigarette smoker	Cancer or heart disease
Flying 700 miles by jet	Accident
Flying 4,000 miles by jet	Cancer from cosmic rays
Living 20 years near a PVC plant	Cancer from Vinyl Chloride
Living 36 hours in Boston or New York	Air pollution

An interesting and comprehensive discussion centred round similar US statistics but expressed in an alternative form is given in Cohen (1983). Here, the average amount by which a person's life may be shortened by a given feature is termed the Loss of Life Expectancy (LLE). For example, Cohen quotes the average life expectancy of a 40-year-old US male as 34.8 years, so that, if he accepts a 1% chance of immediate death, his LLE would be 1% of 34.8 years or 0.348 years. This does not mean that 1,000 such men taking the given risk would all lose 0.348 years of their lives, but that 10 would lose 34.8 years and 990 would be unaffected. A few of the more interesting examples for the present discussion are tabled below:

Table 8.3

The Relation between Various Activities and the Loss of Life Expectancy

Activity	**LLE (days)**
Smoking 10 cigarettes per day	1,600
Working on demolition	1,560
Working as a coalminer	1,100
Being poor	700
Being 7 kilograms overweight	450
All accidents	435
Accidents in the home	74
Occupational accidents	74
Fuel conservation (through reduced ventilation)	24
Fuel conservation (through reduced lighting)	5

The above figures are not precise, since many effects are interactive - air pollution and smoking in combination are estimated to cause many more deaths than indicated by simple addition of the individual effects. There are complicated issues involved in being poor; for example, the inability to pay for adequate lighting can be a contributory factor. Another example is that in the UK it has been shown, from a study of workers made redundant in the 'slimming down' of British Steel in the early 1980s, that unemployment (and consequent loss of income) increased the suicide rate significantly (Haslam, 1988). Cohen's data is now a few years out-of-date and may be imprecise for conditions appertaining to the UK, but it serves to give a very useful perspective.

A fairly recently developed method for estimating the worth of various health procedures takes into account not only how many years longer a patient may live, but also the likely quality of life during those years. To this end, the index of QALY (Quality-Adjusted Life-Year) has been defined. For example, a patient enjoying good health for ten years after a heart transplant will have gained ten QALYs as a result of the operation. A patient treated for leukaemia may also live for ten more years, but it may be considered that because of limitations in lifestyle, loss of well-being and frequent hospital treatments, the quality of life will be only half that of a normal person. The corresponding QALY figure is therefore only five. When the treatment costs are considered, values of procedures in quite different areas of health can be compared. For example, if £50,000 is available for either a heart transplant for one patient or chemotherapy for five leukaemia patients, the cost per QALY would bring respectively,

£50,000 for 10 QALYs = £5,000 per QALY (heart transplant).

£50,000 for 25 QALYs = £2,000 per QALY (chemotherapy).

The leukaemia treatment programme would therefore be assessed as 2.5 times more beneficial per unit cost.

A list of QALYs for many treatments can be set out with corresponding unit costs. The examples below are taken from OHE (1985) and Williams (1985).

Table 8.4

The Cost of Various Medical Operations per Quality-Adjusted Life-Year

Procedure	**Cost per QALY (£)**
Heart transplant	5,000
Kidney transplant	3,000
Hip replacement	750
Pacemaker insertion	700

It is interesting to value in the above terms the enhanced cleanup of Sellafield discharges in relation to the (disputed) ensuing four deaths from leukaemia (see Sections 3.1.3 and 5.2.4). Assuming that each death was equivalent to 50 QALYs, the cleanup, involving several hundred million pounds, cost about £100M per death or £2M per QALY. This corresponds to factors of not less than 400 and 3,000 greater respectively over the QALY valuations for heart transplants and hip replacements in the above table. Clearly, the cleanup money could have been far better spent in improving health in other ways; the equivalent of tens of hospitals could have been built or queues for hip replacements eliminated. (Not only are financial factors involved. In 1988, it was estimated that most patients for hip operations had to wait between 15 and 28 months. The acute pain suffered while waiting often led to mental breakdown and in some cases to suicide. The long waits resulted in at least six suicides per year (Royal College of Nursing, 1988).)

A simple further approach to life valuation can, of course, be derived from actuarial data used by many insurance companies. Figures in the region of £100,000 and more are often quoted (Pearce, 1983): as a 'reasonable' average, in line with similar recent assessment by NRPB (Clarke, 1982), we shall use here a figure of £200,000 as a standard in deriving data for power system economics. (This gives a corresponding level of £4,000 per QALY.) For example, taking a figure of 50 deaths per GW(E)yr from respiratory diseases due to coal-fired power stack discharges, a valuation of 50 times

£200,000 = £10M is obtained, corresponding to 0.1p per kwhr in electricity costs. On the other hand, the Sellafield discharge cleanup figures derived above correspond to £100M per death, which would suggest that a penalty 500 times greater, i.e. 50p per kwhr, should be added to coal-fired power in respect of deaths from stack discharges. Even a tenth of this valuation rate would rule out coal relative to nuclear power. It should be added that these simple calculations do not allow for loss of life quality during a number of years before death from respiratory diseases. For example, it has been estimated (Comar, 1976) that there are 60,000 cases of respiratory disease per GW(E)yr from US coal-fired power plant stack emissions. Even at a low value of 1 QALY for each case of £4,000 per QALY, the cost equivalent of these cases is £4,000 x 60,000 = £240M. This is an immense penalty which must surely rule out coal as a source of power.

Data on damage caused by power systems to flora, fauna and artefacts, is sparse. One source giving estimates of damage caused by Acid Rain is Pearce (1987), who quotes 'some economists' as setting the repair costs of buildings in Europe at £600M per year - say £3M per GW(E)yr: this compares with about £20M per GW(E)yr for West Germany as derived from data of Day (1985). Corresponding figures for US property damage per GW(E)yr may be estimated as about £25M (Barret, 1973). Loss of fish due to Acid Rain pollution in Sweden could cost around £6M per GW(E)yr (Pearce, 1987). Since £1M per GW(E)yr is roughly equivalent to 0.01p/kwhr, the above costs add significantly to the effective costs of power from coal, but not so seriously as for the human health effects described above.

8.5 THE FUTURE UK POWER MIX

8.5.1 Medium-Term Strategies

One factor which determines the contribution that a power system may make to the overall electricity supply is its reliability. This can be considered from several angles. Firstly, the availability of plant operating a proven process depends on the frequency of breakdowns and repair times. Secondly, external factors, such as the effects of miners' strikes on coal supplies, or renewable systems being unable to operate under natural cycles of calm or stormy weather or the state of the tide, may be important. Thirdly, in the long term, supplies of uranium for a 'Once-Through' nuclear cycle could become very costly or politically difficult to obtain: coal-fired power may be curtailed by realisation of the serious effects on health or sudden increases in Acid Rain and Greenhouse effects.

The development of new systems, too, can affect forward planning. At this point in time, the assumption that nuclear fusion can be developed to a successful commercial system in the next (say) 50 years would not appear to be a sufficiently certain basis for inclusion in current planning of future power mixes. The same conclusion can be drawn for other developing systems with the possible exception of wind power, whose contribution to the overall UK power supply could, in any foreseeable scenario, be relatively slight. On the other hand, the PWR is well established overseas and would therefore seem to have a good chance of success if installed in the UK. However, the lessons of Sizewell B as constructed under UK regulations may need careful assessment before a large tranche of PWR stations is constructed. The corresponding delay, while sufficient confidence in eventual success is acquired, may then be reflected in their rate of installation.

The introduction of PWRs, mentioned above, brings an extra flexibility to electricity supply, since the characteristics of their reliability are different from non-nuclear systems. Indeed, such an increase in the diversification of supply is regarded in some quarters (McKerron, 1984) as a stronger reason for a PWR at Sizewell than the case put forward by the CEGB, which was mainly based on costs. The converse strategy of phasing out nuclear power (which would then be difficult to restart) would virtually eliminate diversification and competition to coal.

Overall, therefore, the case for Sizewell B being the first of a series of PWRs, as presented by the CEGB and qualified in the excellent summing-up by the Inspector, appears to provide a reasonable and reliable path for the next few decades of power supply for the following reasons.

(a) Nuclear plant costs are likely to be lower than coal plant costs over this time (for an interest rate of 5%, and assuming no drastic reductions of coal prices).

(b) The environmental impact of nuclear plants is much less than coal plants, even if expensive emission control arrangements are installed in the latter.

(c) In the very long term, cheap fossil fuel supplies in the UK (and the rest of the world) will begin to run out (Pryke, 1987, p.82). It is uncertain how many more 'Belvoirs' or 'Selbys' can be found.

(d) The transmission of power could also be reduced; the need to site coal-fired power stations mainly near to collieries currently incurs losses in overall efficiency because of the higher demand:plant ratio in the south of England. The corresponding possibility of expanding the National Grid, which is near its maximum capacity (Blowers, 1987), could be avoided.

8.5.2 **Long-Term Strategies and the Fate of Plutonium.**

The above analyses of the UK power mix extend only to the next few decades; the nuclear systems discussed, however, are incomplete since the eventual fate of plutonium is left undecided. There are three broad types of strategies which might be followed in the long term.

A. Early Reprocessing.

It is assumed in the CEGB Sizewell cost studies that Thermal Reactor spent fuel would be reprocessed at Sellafield and the resulting wastes treated and eventually disposed. This policy might be pursued in the long term, but low prices of uranium and associated long-term availability could persist, reducing the need and urgency to recover and recycle the plutonium; eventually, the development of a new power process, such as nuclear fusion, may permanently remove any probable use for plutonium. In this case, the reprocessing would have been to no purpose, possibly making disposal problems worse than for the original spent fuel. On the other hand, there are several reasons which could sway a decision towards the continuation of reprocessing, at least for some types of reactor fuel.

(a) Long-term storage of Magnox spent fuel, as at present under water, is susceptible to corrosion and consequent leakage of radioactivity through the fuel cans. Although air storage of Wylfa Magnox spent fuel appears promising, the closure of most Magnox stations in the next decade or two would probably make it not worthwhile to switch over to a new procedure. Moreover, the disposal of Magnox spent fuel as such could be more difficult to achieve with adequate safety than Oxide spent fuels (see Section 4.1.8.3).

(b) A U-turn in the commercial commitment to reprocess Japanese spent fuel would be embarrassing and costly.

(c) The THORP plant is at an advanced stage of construction, probably going 'active' in the near future. It would be difficult not only to 'mothball' the plant for reopening in (say) 50 years, but also to recruit the large numbers of staff of special skills necessary to start the plant up again and operate it. Scientists and engineers would indeed be very wary of rejoining a project shut down for what many would consider as trivial political reasons.

(d) Current system studies by the Nuclear Energy Agency suggest that recycling of already separated plutonium in MOX in Thermal Reactors would be cost effective and not affect the future use of the residue in Fast Reactors. The studies also indicate that separation of plutonium in existing reprocessing plants for recycling in MOX fuel in Thermal Reactors may be marginally economic, but the special construction of new reprocessing plants for such recycling does not seem justified on cost. In practice, the Federal Republic of Germany is already using MOX in reactors; Belgium and France, too, are planning the same and possibly Sizewell B will follow at some stage. On the other hand, plutonium recycling through AGRs is less economic than in PWRs, so is less likely to occur. A further source of MOX fuel could arise as plutonium needs for weapons decline.

B. Early Disposal of Spent Fuel by Deep Burial.

From the discussion and data of Appendix 4.6, the cost of excavation of very deep boreholes, down which spent fuel sub-assemblies could be emplaced, is small in relation to other fuel cycle costs. The operation would appear feasible, since similar oilfield technology and engineering should be adaptable for the safe lowering of the fuel and eventual sealing up of the boreholes. The operations are simpler than the above early reprocessing route, so the overall disposal cost is therefore unlikely to be greater than the Sizewell costs put forward by CEGB (disposal costs for HEW relative to the corresponding electricity value are fairly small, as discussed in Section 8.3.3). The overall cost of such a 'Once-Through' system (as described earlier in Chapter Seven) should therefore be at least as advantageous relative to coal as the incomplete nuclear schemes put forward by CEGB. However, the irretrievable disposal of the plutonium in the spent fuels, which would deny to future generations the option of operating Fast Reactor systems with the equivalent energy content of tens of billions of tonnes of coal, is a most unlikely decision to be taken by any UK Government in the foreseeable future.

C. Long-Term Storage of Thermal Reactor Spent Fuel.

The AGR and PWR fuels are sealed in either stainless steel or zirconium alloy cans. Cooled by a flow of air, such can materials are satisfactorily stable for several decades (see Section 3.2.6). Reprocessing and waste disposal would then involve handling materials of activities appreciably lower than currently carried out, resulting in significant cost savings. Overall, therefore, the additional costs of long-term storage before reprocessing or deep burial would probably not result in an overall increase in system costs compared with early reprocessing.

There are clearly 'pros and cons' for each of the long-term individual strategies A, B and C above, all of which would have lower costs than a coal-based power system. A flexible compromise, which would reduce some of the objections to individual schemes, could be to continue reprocessing on the present scale, accommodating excess fuel arisings over reprocessing capacity by air cooling appropriately canned fuel in long-term storage. This 'buffer' store of fuel could be treated according to the requirements of fuel cycles of the types outlined in Chapter Two, as considered appropriate in the next century. On the other hand, the sale or loan of plutonium stocks might also be found expedient as a means of smoothing out the peaks of reprocessing overseas fuel.

8.6 SUMMARY

The annual average demand for electricity in the UK has been increasing slowly in the last decade or two. Superimposed on this average requirement are fluctuations, which are seasonal (from changes of climate), daily (due to industrial and domestic activities) and brief daytime peaking (domestic appliances). Extrapolating this demand sequence for the future is known from past experience to be an uncertain procedure. The most recent authoritative forecast is that put forward by the Department of Energy at Sizewell, which broadly assumes that the present slow increase in demand will continue.

The cost of electricity from the National Grid can affect the total demand according to its competitiveness with alternatives such as various forms of conservation. A basic energy requirement is therefore to satisfy the demand pattern in the most economical manner; in recent years, however, other considerations such as the effects of routine discharges and accidents have also become important. The situation is further complicated because judgements on generating costs are affected by their occurrence at different points in time and from different plant sizes. The accepted procedure for the correlation of such costs (Discounted Cash Flow or DCF), with allowances for various plant sizes, was applied by the CEGB for presentation at the Sizewell Enquiry. The figures showed an advantage for nuclear over coal power systems, especially for the PWR. Objectors raised several issues which suggested that these advantages were not as clear cut as presented. On the other hand, it was felt in some quarters that, apart from cost aspects, the PWR could be justified in increasing the diversification of generation methods. Though only partially accepting the CEGB case, the Enquiry Inspector was generally in favour of the points made in the proposal. The go-ahead for Sizewell B has now been approved by Parliament. It is interesting to note that,

(a) the Government found it expedient to say, in the same announcement as the approval for the go-ahead for Sizewell B, that two new coal-fired stations would be built. This is clearly inconsistent with the CEGB conclusion that further coal-fired plants to replace existing base load plants would be uneconomic (Section 8.3.3). Over their lifetime, these stations will condemn many thousands to an early death;

(b) there were no protesting cries from the above Sizewell objectors concerning further overcapacity from the new coal plants. Clearly, the net result will eventually be that even more old coal-fired plants are 'pensioned off'!

The Sizewell Enquiry did not address several issues.

(a) It only compared plant costs, but not environmental issues, e.g. there was no comparison of discharges and operational waste management from nuclear and coal power systems, especially to the same environmental standards.

(b) A particular aspect of waste disposal subject to much Enquiry time was that of decommissioning nuclear plants. Unfortunately, there was no corresponding discussion of the decommissioning of the immense spoil and fly ash dumps remaining from coal power operations, nor of the costs of averting subsidence above mines. Judged on the same basis as these coal waste heaps, it would seem unnecessary to remove shutdown nuclear plant completely.

(c) The fate of plutonium was more or less left undecided, largely because of the relatively short period which the forecasts covered.

Analysis in Section 8.4 of the effects of human activities and experiences, derived mainly from Cohen (1983) and O'Brien (1986), provides a basis for estimating values of health and other environmental detriments relative to plant costs. At a conventional life valuation of £0.2M, deaths from stack discharges add a significant increment to coal power costs; assessment with a life valuation equivalent to even one-tenth of that used for Sellafield pipeline cleanup (Sections 4.1.3 and 6.2.4), however, would raise coal power costs to quite unacceptable levels. Though these quick comparisons are necessarily very rough, they reveal glaring inconsistencies between standards applied to nuclear and coal power operations. One way of bringing such assessments into line is to treat the detriments by modern techniques which compare the value of different health procedures. Using some of the figures from such techniques as quoted in Section 8.4, we may extend the CEGB comparison at Sizewell to deduce that the installation of Sizewell B relative to a coal-fired plant of similar size (not necessarily at Sizewell) could each year save about 50 deaths and yield savings which could pay for either 6,000 hip replacements, 1,500 kidney transplants or several new hospitals. The eventual increase of the nuclear component to 50% of UK power would correspondingly save many thousands of deaths and pay for substantial improvements in the National Health Service. Acid Rain damage to the environment other than man appears to be less important on cost grounds. (It might be argued that savings (say) in armaments could pay for similar health improvements; however, this introduces other factors such as national security. On a straight health comparison, nuclear power plus improved national health must always have higher priority than coal-fired power.)

Concerning the fate of plutonium, the discussion in Section 8.5 suggests that if the CEGB assessment is extended to follow possible long-term strategies, i.e. a Once-Through Cycle, recycling plutonium through Thermal Reactors, Fast Reactors or a combination of the two, the cost advantage of nuclear over coal would be maintained, if not increased. NEA studies, in fact, suggest the total fuel costs in Once-Through Cycles to be less than reprocessing cycles (Jones, 1987) by about 10%, according to the discount rate assumed. In the medium term, the issue of reprocessing to recover plutonium is of lesser importance than the choice of nuclear versus coal, since reprocessing is relatively innocuous environmentally compared with coal power operation (Chapters Six and Seven) but, on the other hand, there is no urgent need to use plutonium in future fuel cycles. A compromise strategy is suggested here of maintaining reprocessing at a level matching the throughputs of existing plants including THORP, but not building new plants to expand overall capacities. The residual spent fuel arising from the limited number of years of operation of Magnox reactors could then be reprocessed. At the same time, stores could be built to store excess arisings of fuels with cans of appropriate stability for long-term air cooling, before eventual reprocessing or disposal.

Chapter Nine

POLITICAL ASPECTS OF UK POWER

9.1 GENERAL

Since the development of nuclear power began shortly after the Second World War, the number of organisations associated with various aspects of electricity production has increased considerably. All of them have some influence on present and future UK power, so in the next five sections a brief description of the group of organisations for each type of power generation is presented. The two succeeding sections (9.7 and 9.8) describe the UK media and public responses to power issues. Sections 9.9 and 9.10 outline the attitudes of Government and Opposition parties. Section 9.11 begins with the 'Economy of Truth' or the presentation of 'facts', then analyses the driving forces behind the protagonists, comparing the 'S-J Cycle' (Self-Interest and Justification) with Parkinson's Law. A short section (9.12) then discusses power situations in other countries and their relevance to the UK. The final section is a synopsis of the chapter.

9.2 THE 'NUCLEAR CLUB'

The obvious technical attractions of nuclear power, coupled with the desire of the Government to break the monopoly of coal in UK power supply, gave a considerable impetus to the setting up of nuclear power organisations, which collectively are often referred to as the 'Nuclear Club'. In 1954, the United Kingdom Atomic Energy Authority was created (UKAEA). In the following year, the go-ahead was given for the construction of twelve nuclear reactors on the premise that the associated electricity costs would be comparable to those of coal. Five consortia of companies were formed to build these Magnox reactors; over the next decade or so, however, the multiplicity of variants on the original prototype designs led to increased construction times, a situation which was repeated when AGRs replaced the Magnox system. Accordingly, the constructors were eventually all grouped together in the period 1973-6 into the National Nuclear Corporation (NNC).

In the 1960s, the Production Group of the UKAEA was split off to become British Nuclear Fuels Limited (BNFL). All the fuel cycle operations of Thermal Reactors were carried out by the new company, though responsibilities on wastes, other than LLW, extended only to storage, not subsequent disposal. The UKAEA still controlled the development of prototypes of new reactor concepts and also Fast Reactor fuel reprocessing. Though the company structures of NNC and BNFL provided the organisations needed to compete with the large nuclear groups overseas, it also created monopolies in the UK. The lack of competition meant that increased costs were passed on, from BNFL and NNC to the CEGB, who then added them to their customers' electricity bills.

In 1982, the development of disposal methods for LLW and ILW became the remit of NIREX, whose personnel were formed from the staff of BNFL, UKAEA and the electricity producers, CEGB and SSEB. Later, in 1985, NIREX was reconstituted as UK NIREX Limited, with all shares held by the above partners and the UK Department of Energy. The current problems of NIREX, with respect to the 'NIMBY' attitude of the public, are taken up later in Section 9.8. It suffices to say here that early NIREX approaches to radioactive waste disposal, with a leaning towards the application of the best technical solution on grounds of cost and safety, have given way in the late 1980s to a political slant, in which the public are invited to discuss how radioactive waste should be managed (NIREX announcement, November 1987).

In the 1980s, nuclear reactor issues have been dominated by the Sizewell Inquiry and Chernobyl. The CEGB, whose background is further elaborated in Section 9.5, had decided to switch from the British-designed AGRs to the Westinghouse (US) PWRs: the first of these was planned to be constructed at Sizewell in Suffolk. During the Inquiry, the prospect of a smooth expansion of nuclear power received a severe shock with the news of the incident at Chernobyl: even so, the PWR proposal was given the go-ahead.

At the present time, with the continuing discovery of low-cost uranium ores, the consequent decreasing urgency to install commercial Fast Reactors and the absence of the need to develop new designs of Thermal Reactors, the importance of the role of the UKAEA has diminished. In fact, recently, in July 1988, the UK Government announced severe cuts in expenditure on Fast Reactor research and development, including withdrawal of financial support at Dounreay for the Prototype Fast Reactor in 1994 and the associated fuel cycle plants in 1997. It seems likely that only one Commercial Demonstration Fast Reactor will be built in Europe, if at all.

Much of the UKAEA's attention has turned to the 'Back End' of the nuclear fuel cycle - research in support of NIREX, operational waste management and the decommissioning of reactors, as exemplified by the exercise launched to remove the Windscale AGR and return its site to a 'green field' condition. Research on the disposal of HLW has remained with the UKAEA and DOE; this is, however, confined to 'generic' investigations, i.e. of general types of site conditions and not of any specifically selected site.

With the likelihood of application of Fast Reactors receding in time, the need to recover plutonium for their fuels from spent Thermal Reactor fuels has diminished, so that the justification for the associated reprocessing at Sellafield is less clear: a large part of BNFL's operations are thus in jeopardy. As mentioned in Section 8.5.2, alternatives to early reprocessing include disposal of spent fuel or long-term fuel storage, say for 50-100 years, in case Fast Reactors are needed. It is hardly credible that spent fuel should be 'thrown away', as it represents a fuel potential for future generations equivalent to tens of billions of tonnes of coal. On the other hand, deferring reprocessing for 50 to 100 years requires a considerable investment in storage facilities and the winding down of BNFL's reprocessing plants with a corresponding loss of the skills of thousands of experienced staff. It is uncertain, too, whether such long-term stores for spent fuel would necessarily be built by BNFL or NNC. Restarting in the distant future would be a lengthy and costly exercise with the likelihood of replacing much 'old-fashioned' plant: given the probable continuation of reprocessing in France, West Germany and Japan, it would seem difficult for a future BNFL to re-enter the market successfully. On the other hand, a future Government may seize the opportunity to introduce competition in reactor building and fuel supply from Europe and Japan, after the manner of the UK motor car industry in the last few decades.

During the long time interval prior to Fast Reactors becoming necessary, plutonium may be recycled through Thermal Reactors. The case for this appears marginal (Section 8.5.2), appearing economic only if the capital cost of the reprocessing plants has already been paid. The justification of further BNFL reprocessing plants after THORP could therefore be considerably affected by the availability of overseas funding.

9.3 BRITISH COAL AND THE MINERS' UNIONS

Just after the Second World War, the UK coal industry was nationalised, since when it has been managed by the National Coal Board, renamed in 1986 as the British Coal Corporation (often abbreviated to British Coal). The workforce has dropped as improved mining techniques have been introduced and the demand for coal has fallen. The main union representing the workforce is the National Union of Mineworkers (NUM). A confrontation with the National Coal Board in the early 1970s led to a strike, which resulted in victory for the NUM and effectively the downfall of the UK Government. A further strike in 1984, however, brought defeat for the NUM, a substantial minority of miners continuing to work and forming a breakaway union, the Union of Democratic Mineworkers (UDM). Apart from the actions of UDM miners, other reasons for the NUM defeat were the availability of large stocks accumulated by the National Coal Board before the strike and the provision of power from other sources, mainly nuclear and oil.

The NUM has now lost its dominating influence on coal supplies, so that the political driving force to develop alternative power systems has decreased. Nevertheless, the policy of the present UK Government is still to increase the diversity of power systems and hence the reliability of the overall national supply. The future of the British coal industry is therefore full of uncertainty, with reduced demands for coal power, the threat of cheap imported coal and the possibility of some form of privatisation.

9.4 FUSION AND RENEWABLE POWER DEVELOPERS

The basics of fusion power were described in Section 1.4.2. Large research machines are required in order that the product of the concentration of 'fusible' nuclei and their residence time in the reaction zone creates a meaningful number of reactions for measurement. The resources of large countries such as the US or USSR, or groups of states as in the EEC, are therefore required. The largest machine now operating in the European Atomic Energy Community (EURATOM) is the Joint European Torus (JET); this is situated on the UKAEA site at Culham, near Oxford. Other large machines are the TFTR at Princeton, US, and JT-60 in Japan. Current research suggests that, in order to achieve continuous power production at a commercially worthwhile level, machines will have to be much larger still. As a consequence, the EEC, Japan, the Soviet Union and the US have agreed to start joint planning work for an international fusion experiment. The Max-Planck-Institut für Plasma-Physik (IPP) at Garching near Munich is the proposed technical site. Given project approval, construction of this International Thermonuclear Experimental Reactor (ITER) could start about 1993; if successful, ITER could provide the basic information for the design of a demonstration thermonuclear reactor.

Of the renewables, tidal power in the UK is in an early stage of development, so that the corresponding organisations for its promotion are not yet well defined. To date, the Severn Barrage Committee has used the consulting engineers Binnie and Partners for feasibility studies. Wind power looks the most attractive currently among the renewables: availability of 90-95% for UK conditions has been predicted (Milborrow, 1985). Three major companies, Howden, McAlpine and Taylor Woodrow, have entered the field with units in the megawatt range; most of the sales of commercial units, however, have been in the 60 to 300 kilowatt range for application overseas.

Various wave power devices have been outlined in Section 1.4.3. Of the large number of possible systems, the three described there are capable of scale-up to outputs significant to the National Grid. The SEA Clam is being developed by several companies grouped as Sea Energy Associates, based on pioneering work at Lanchester Polytechnic: the Lancaster Flexible Bag was invented by Professor French of Lancaster University and the NEL Oscillating Column is the brainchild of the National Engineering Laboratory. Two Norwegian coastal wave systems of possible application in the UK and therefore worthy of mention are,

(i) the Tapchan Project, which focuses waves in a convergent rock gully, spilling over the channel sides into a reservoir feeding a turbine, the exhaust draining back to sea, and

(ii) the oscillating water column of the Kvaerner Brug company, which is recessed into a cliff, the water fluctuations forcing air through a turbine.

The UK Department of Energy, which sponsors research and development into renewable systems, has recently issued a useful review of the 'state of the art' (DoEn, 1988b).

9.5 SUPPLIERS OF ELECTRICITY

In 1957, the Central Electricity Authority was dissolved and the Electricity Supply Industry (ESI) in England and Wales was set up, comprising the following statutory bodies:

(i) the CEGB, responsible for the generation and transmission of electricity in bulk through its National Grid,

(ii) the twelve Area Boards, which receive bulk supplies from the CEGB for distribution through their own networks, and

(iii) the Electricity Council, where the general policy of the ESI is discussed. The thirteen Boards had to consult the Council in certain matters, particularly their capital investment, but the Council had no powers to direct or control them.

In Scotland and Northern Ireland, electricity supply is the responsibility of Boards, independent of the ESI Boards.

As at 1988, the CEGB owns and operates the power stations and the National Grid. The system has a net capability of about 60 GW(E): the disposition of the largest power stations is shown in Fig. 9.1. Because of the early dominance of coal-fired power, many of these stations are sited along large rivers and near coalfields. The demand pattern is somewhat different geographically (see Fig. 10.6), thus it is desirable to site new nuclear stations in southern England to reduce transmission requirements.

Generation has been split into five regions for organisational purposes - the South Eastern, the South Western, the Midlands, the North Eastern and the North Western regions. Grid control centres within these regions are controlled by the National Control Centre in London to ensure system efficiency and security by arranging power transfers in advance.

In addition to the Regional organisation there are three Divisions - the Generation Development and Construction Division at Barnwood near Gloucester, the Transmission and Technical Services Division at Guildford and the Research Division at Leatherhead, Marchwood and Berkeley. Among the research topics carried out by the CEGB are programmes on air pollutants and their effects and (jointly with British Coal) pressurised fluidised bed combustion - the largest test facility in the world for this is at Grimethorpe, South Yorkshire. Further, the CEGB carries out research and development to keep abreast of possibilities in the renewables and CHP. For example, it considers wind power sufficiently promising in some locations to plan ten sizeable machines by 1990.

Under the Government proposals to privatise electricity generation and supply, the CEGB would be split into two privately owned components - little 'G' and big 'G'. The intention is to break the present CEGB monopoly and introduce competition into the electricity system.

9.6 THE 'GREENS'

'Green' groups, such as the Friends of the Earth (FOE) and Greenpeace, have increased considerably in prominence in the last decade or so. National groups of similar persuasion have linked together to form comprehensive international organisations.

Initial ecological issues, such as opposing aspects of whaling, had fairly simple commercial connotations, so that there was an emotional appeal and also a straightforward rationale with which to gain public support. The need to retain such support and associated interest has led to a natural expansion into other issues. Of these, the obvious growth topics are those which readily attract public concern and so will be raised to a 'high profile' by the media. For example, the present Director of FOE is well aware that 'television is much more powerful than schools, churches or the family in terms of creating a set of values' (Porritt, 1984). It is not surprising, therefore, that one central theme which has developed is their opposition to nuclear power. The supporters of the 'Greens' have, by and large, strong emotions and views, with extraordinary optimism on costs of renewable systems. At the same time, they refer scornfully to the rosy picture of nuclear power, painted on more substantial grounds in the early 1950s.

In early attitudes, the 'Greens' claimed that nuclear power was dearer than power from coal. However, when FOE did a cost comparison of the two systems a few years ago, the strange conclusion was drawn that 'coal power was the cheaper if only miners would forgo any wage increases for five years'! It is now admitted (Porritt, 1984) that 'if one's goal is maximisation of economic output ... then nuclear power ... (is) ... rational.' The broad approach now adopted by the 'Greens' is that coal-fired power should be reduced (because of the Greenhouse effect) and that nuclear power should be abandoned because 'the UK should set an example in reducing nuclear proliferation'. It is assumed that conservation will drastically reduce UK power requirements and that renewables will provide such power as is not forthcoming from CHP (with coal as the energy source). It is not clear what industry and domestic appliances will do when the wind, waves or the sea decline to co-operate! Costs are deemed unimportant - 'Of course there are problems and of course it's expensive' (Porritt, 1984) - yet at the same time Fast Reactors are ruled out on cost (Flood, 1987).

The manner of presentation of 'Green' views is compared with those of the Nuclear Club and the Media in Appendices at the end of this chapter.

9.7 THE MEDIA

9.7.1 General

Almost by definition, the media exist to transmit information to the public. An account of their development is given in Appendix 9.3, mainly abstracted from Curran (1988). In order to attract wide interest and thereby to obtain good financial returns, there is a natural inclination towards sensation and a consequent aversion to dull factual reporting. Indeed, one editor has been known to declare that, if there were a serious setback in the national economy and also the birth of a three-legged sheep, he would highlight the story of the sheep! For our discussion on the attitudes towards the various power systems, it is convenient to define the media as comprising three main groups.

A. The 'tabloids' are small-page 'popular' newspapers which cater for stimulation of readers through sex, horrors, scandal, sport, etc. Attempts to educate the reader are simplistic and brief.

B. The 'holoids' (holier than tabloids) or 'quality' press, with the exception of weekly journals, are usually printed on larger pages than the tabloids. They purport to educate their readers and keep them fully informed. A measure of sensation is, however, introduced to boost sales, though this is usually done in a more subtle manner than in the tabloids.

C. The 'teloids' are the television companies, currently under the administration of the British Broadcasting Corporation (BBC) and The Independent Broadcasting Authority (IBA). There are plans to introduce several more TV channels in the early 1990s.

9.7.2 Tabloids and Holoids

The general approach by newspapers to the power debate is well illustrated in the comments below by two leading journalists. David Fishlock of the Financial Times recounts how the editor of one British newspaper told the Chairman of UKAEA, 'I'm not against nuclear energy - but your industry does make good headlines'. Fishlock continues:

> This is because newspaper readers have an immense capacity for shock and horror stories. The readiness of at least some sections of the newspaper-reading public to believe the worst that can be invented about the nuclear industry seems to be matched only by the uncritical credulity with which they accept claims for free energy from waves or wind. However, when faced with a more critical examination, 100 days of testimony by opponents of nuclear energy before a public inquiry in 1977 failed to persuade Judge Parker to accept a single one of their 17 separate arguments against the Windscale project for managing nuclear wastes. ... The nuclear industry is feared by those who want radical political change. Nuclear energy has immense capacity for promoting political stability. That, above all, is what its most serious opponents fear most, and why they so earnestly try to persuade the public to reject all things nuclear.
>
> (Fishlock, 1984)

An article by James Wilkinson, the science correspondent of the BBC, includes:

> What science correspondents write is to a degree influenced by the knowledge of what their newspaper will publish, both in terms of what the paper considers would interest the readers and, of course, in terms of the politics of the newspaper. Sometimes articles about nuclear matters are written by general reporters with no background knowledge of the subject.
>
> (Wilkinson, 1985)

A personal experience by the author of this book illustrates this point. During a discussion with the editor of a holoid on erroneous facts in nuclear power articles written by his science correspondent, the editor protested, 'He has the whole field of science to cover and so has no time to understand what he's writing about'!

Covering pressure groups, Wilkinson acknowledges:

> They have been highly successful in focusing public attention on their views through the media. In recent years, Greenpeace particularly has been making considerable impact. It makes little secret of the fact that it wants to get rid of the nuclear industry altogether. Their anti-whaling campaign has had a 'halo' effect which puts the organisation in a favourable light as far as the public is concerned and therefore lends more credibility to its anti-nuclear campaign. Again, Greenpeace had almost unbelievable luck when they were the first to proclaim the discovery of a radioactive slick offshore at Sellafield in 1984, announcing it ahead of BNFL.
>
> (ibid.)

Wilkinson suggests that it would be quite wrong for the media to suppress the reporting of Greenpeace stunts and that the nuclear industry should set up counter-stunts, citing the success of the CEGB-organised train crash to show little damage resulted to a nuclear transport flask in a severe accident. This demonstration was covered by both the BBC and IBA. He also cites the influence of competition between newspapers:

> The first editions of newspapers are rushed around Fleet Street to competitors at midnight so that each paper can see what their rivals are printing. If one newspaper has a story about (say) radioactive waste which is headline-grabbing ... it is picked up and printed the same night by others. So what began as an exclusive story in one paper may find its way into all the others by morning.
>
> (ibid.)

Asked what could be done about misleading articles, Wilkinson said that for major factual errors a correction would usually be printed. Otherwise there is the possibility of recourse to the Press Council, recruited from within the newspaper industry.

With respect to Wilkinson's last remark, Curran (1988) suggests various reforms of the Press Council and, where complaints are upheld, these should receive prominence equal to the offending articles (see Appendix 9.3).

9.7.3 **The Teloids**

In the setting up of the BBC, it was given a charter stating specifically that, 'due impartiality is (to be) preserved ... as respects matters of political or industrial controversy or relating to current public policy' and 'all expressions of their own opinion as respects matters of political or industrial controversy or relating to current public policy' must be excluded from the programmes by the producers and all concerned.

Although the Independent Broadcasting Authority (IBA) does not itself make programmes, it is ultimately answerable to Parliament and the public for everything it transmits. The Broadcasting Act requires the IBA, 'to ensure ... a proper balance of information ... accuracy in news, due impartiality in matters of political and industrial controversy, etc.'

The three main political parties have expressed widely different views on UK nuclear power, from continued expansion to complete shutdown (Sections 9.9 and 9.10). In spite of this, programmes from both the existing teloids highlight alleged shortcomings of particular aspects of nuclear power. Two general approaches are,

(a) historical reviews attempting to show how forecasts and promises have not been achieved, thereby reducing public confidence in current Nuclear Club statements, and

(b) 'What If' programmes, sensationalising the consequences of low-probability disasters.

With respect to the calculated flouting of teloid charters by distorted reporting, the increased powers proposed for the Broadcasting Standards Council, to be set up in the next year or two, whereby corrections and apologies must be screened at the same time as the offending programme, could with advantage include long-term corrections to counteract long-term bias. Indeed, the Deputy Director-General of the BBC has called for a determined effort by the media towards self-regulation in order to avoid impositions restricting their legitimate activities (Birt, 1988).

It will be interesting to see what forms of Charters are drawn up by the Government when there are perhaps a dozen TV channels. Will the multiplicity be regarded as adequate safeguard against political bias, so that viewers can choose the channel of their political preference, just as readers of newspapers are free to buy the paper to suit their 'political palates'?

Currently, however, the political attitudes of a small number of media moguls and journalists can 'govern by chat show or sensation' and affect important aspects of the nation's future. In the case of nuclear power, the many lifetimes of effort by technologists who 'answered the call' by the nation after the Second World War, may thus be discarded. Perhaps some were attracted 'by the money', but the throwing away of the fruits of such extensive expertise makes one wonder if the nation could save a great deal of expense by not educating young people but sending them straight into journalism.

Some instances of biased reporting by the media are given in Appendix 9.4 with illustrations in Figs 9.2 and 9.3.

9.8 'NIMBY' AND PUBLIC PERCEPTION OF RISKS

Lee (1985) gives a concise explanation of the perception of risk by the public. 'When a hazardous technical activity or feature of the environment is evaluated by a scientist or engineer, we speak of "risk assessment": when a similar evaluation is made by a non-expert member of the public, the term used is "risk perception". Fundamentally, there is no difference. Both groups rely on their knowledge of past events to make predictions about future events.' Some examples of this are quoted in a recent report by O'Brien (1986) where the studies reviewed show that individuals judge an event to be more likely (than its actual probability) if instances are easy to recall or imagine. Personal experience is a powerful influence on perception. Loss of a relative or friend through leukaemia is likely to cause an individual to exaggerate the incidence o-the disease in the community. Over-estimation of the likelihood of rare events is obviously related in part to the information and news generally available to the public. This question was analysed by Combes (1979) who concluded that media coverage was biased towards the more sensational events threatening life: 'accidents, violence and disasters sell more newspapers and increase television ratings than heart disease and stomach cancer'. It will be interesting to note the public reaction when large wind power devices are erected in various sites, particularly in south west England, as announced by the CEGB in 1988; will they believe the CEGB assessment (Milborrow, 1985) that deaths from breakup of windmill blades are only as likely as being struck by lightning?

The issue of nuclear waste disposal is an extraordinary illustration of the effect on the public of the manner of presentation by the media. Lee (1985) comments,

> It is certainly true that ordinary people's knowledge of nuclear waste management is largely based on investigative journalism about Sellafield, presented on television. The latter wilfully confuses nuclear power for the production of electricity with nuclear weapons or, conversely, keeps asking how many jobs it will create. Emphasis is placed on considerations like the quality of life for future generations instead of comparing the marvellous safety record with the appalling historical carnage in the coal industry.

On the other hand, he also comments that,

> scientists and engineers are also selective, choosing to include in their evaluations those variables that are clearly quantifiable, e.g. money and mortality. They also arrive at conclusions that are compatible with their employment more often than can be accounted for by coincidence. Not because they deliberately distort or select data, but because they choose and are chosen by their employers and this sets their frame of reference.

Later, he makes the observation that he does not think,

> that the press has very much influence on public opinion ... people in that area (Sellafield) have a kind of latent concern about it (radiological injury) and are naturally worried. The Press knows that people will buy their newspapers if they feed that concern back to the public. There has been a systematic study in the US that showed that the reporting of homicides and other violent and dramatic hazards leads to an exaggeration, on the public's part, of the importance of these risks by comparison with others. I would argue that the press report homicides precisely because the public has an enhanced interest and exaggerated perception of their frequency.

An interesting conclusion which Lee draws from an analysis of simple questions on nuclear power, is that,

> The pros and antis do not differ significantly on knowledge. The only trend is that the pros seem slightly more aware of the relative contribution to the radiation levels of, for example, medical X-rays and the antis overestimate the official figures of bomb-test fall-out and radiation levels near nuclear power stations. The particular significance of this finding is that the belief held in the nuclear industry, that people's attitudes will become more favourable if they had more knowledge about nuclear power, is clearly false.

A particular aspect of the above observations concerns the siting of nuclear waste disposal facilities. With coal-fired wastes, the arisings are so huge that there is no option but literally to dump them close to their source. Moreover, there is no question of getting them into the eye of a television camera! On the other hand, nuclear waste volumes are so small that not only are there many choices for their treatment and location of disposal, but it is possible to place substantial barriers between the wastes and possible contact with man. However, since there are few advantages to inhabitants local to such disposal, there is a resentment at receiving the preponderance of the apparent detriments when the rest of the country accepts only the system benefits. The consequent strong opposition to new nuclear waste disposal sites has been dubbed 'NIMBY', the 'Not In My Backyard' syndrome. Curiously enough, on the other side of the fence, nuclear organisations have, in the past, also been reluctant to offer operational sites for disposal of wastes, since the consequent public inquiries are extremely time-consuming and focus attention on their 'patch'. NIMBY is truly universal! The announcement by BNFL in the autumn of 1987 concerning a sub-seabed repository off Sellafield is a big change in policy, possibly stemming from a need to create further work after THORP. The rationale for siting a repository deep underground and yet with the claimed facility to retrieve the waste seems a trifle odd. If the waste is readily retrievable, it may not be under the same 'wet' conditions applicable after 'final' disposal, so that little demonstration of satisfactory 'final' disposal can be claimed. A more satisfactory arrangement would appear to be to store the waste in a Dry Repository (see Section 4.1.8.2): the storage period would demonstrate satisfactory water flows near the tunnel walls and the system could be easily converted to the 'final' disposal mode at a suitable point in time. In the above context, it is interesting to note the subtle changeover in the past few years by the Nuclear Club from the word disposal (which implies losing control) to storage (which implies both control and retrievability).

The decision by the UK Government in 1988 to cut Fast Reactor development severely will increase unemployment appreciably in some areas. Dounreay is one of these, so that the driving force of job scarcity will probably overcome the local 'NIMBY' factor. Both waste repositories and long-term spent fuel stores may therefore quite possibly be sited at Dounreay. If reprocessing dwindles at Sellafield, eventually the same sequence could also follow there.

From time to time, the question of compensation to individuals and/or communities, in respect of disturbances incurred in the broader national interest, has been considered - see, for example, the discussion in Comm. En. Env. (1981). Examples of such compensation are the allocation of a proportion of oil profits to the local authorities in Shetland and the rate rebate allowed in France in areas where nuclear plants are built (Blowers, 1987). The problem is exacerbated in the case of nuclear waste disposal, since the hazard is in reality virtually nonexistent, but distortions in the media can nevertheless create real difficulties in such matters as selling local produce (which the public might believe to be contaminated), reduced house prices in the disposal neighbourhood, etc.

Before ending this section, it is worth noting that other industries suffer from events raised to high profile, e.g. the food and drug industries. Here, rare accidents with (say) a new drug are highlighted which are often far outweighed by its benefits generally. This sensationalising can result in its withdrawal from the market. Such a sequence begs the question as to whether research is worthwhile in such cases and whether the Government should pay damages for isolated mishaps from processes generally approved as being of overall benefit to the nation.

9.9 THE GOVERNMENT

Within Government bodies, there is an understandable reluctance to make positive decisions in the field of nuclear power, particularly as regards waste, when there is little possibility of 'kudos' but a good prospect of 'flak'. It is tempting to put off the 'evil day' with a statement such as 'we are doing research'. Such prevarication leads to many inconsistencies. Research groups can spend ever-increasing sums as the Government 'throws more money' at the problem so as to appear to be doing something. The House of Lords Select Committee recently joined the chorus for 'a Rolls-Royce' solution to nuclear waste. Great news indeed for the Treasurers of Departments! But does the lady who has painfully waited for years for a hip replacement prefer such a 'Rolls-Royce' to lots of 'Morris Minor' hospitals? What is the point of new Rolls-Royces when the Austin Seven pits and trenches already filled at Dounreay and Drigg cannot be similarly upgraded? Further, why should nuclear waste be surrounded by sophisticated multiple barriers on disposal when no other toxic waste is so contained?

One response of UK Governments to public concern has been to set up or reactivate a Royal Commission. The Royal Commission on Environmental Pollution issued a report (RCEP, 1976), one of the chief conclusions of which was that, 'there should be no large-scale expansion of nuclear power until a satisfactory solution had been found to the problem of long-lived radioactive wastes.' However, there was no clue given as to how satisfaction was to be determined. From our comparisons in earlier chapters, relative to the alternative long-lived toxic effluents from coal-fired power, both radioactive and non-radioactive, it is clear that there are already several more than adequate methods of nuclear waste disposal. The safety factors available with these are so large that the scale of the nuclear programme is quite irrelevant.

Another recourse of Government Departments in the recent past has been to institute public inquiries. These have become so long and have generated such mountains of paper as to dull the interest of the public and make it impractical for anyone, not knowing the facts initially, to comprehend the issues. The Sizewell Inquiry lasted several years, recommending the go-ahead for a PWR, the first in an expansion proposed by the Government to provide a 50% nuclear component of UK power in the next few decades (Section 8.3). As a sop to protestors, the Inspector suggested that an investigation be put in hand concerning the management of LLW from the reactor. When Parliament approved the construction of the PWR, it applied a political touch by announcing that two new coal-fired power stations would also be built. Since there will be enormous problems with the associated wastes, both gaseous and solid, will there be inquiries in the vicinity of these stations? Will a study of clusters of cancer and respiratory diseases be made round coal-fired stations generally in the UK? It would be interesting if such a study produced negative results; the solution to possible clusters round nuclear sites could then be merely to build taller stacks so that, as for present coal-fired stations, health problems would be spread uniformly over the UK and the Continent.

The impending privatisation of the power industry will introduce more competition between power systems. It may be noted that despite a 60% increase in output per UK miner since the 1984 strike, corresponding output from the admittedly easier extraction conditions in Australia and the US is four times higher. Foreign coal may therefore be used as fuel for UK power in the future, but what weighting will the Government put on its quality and consequent discharges on combustion?

The effect of privatisation on nuclear power (DoEn, 1988a) - and eventually, too, on fusion, which also requires large generating units to achieve cost advantages of scale - has been mentioned in the last chapter. The high capital costs could render returns unattractive to investors at normal market interest rates. This characteristic favours control of nuclear plant by large national organisations - the opposite of Conservative Party dogma of competition in the private sector between a multiplicity of companies. The setting up of 'Big G' and 'Little G' is clearly an attempt at a compromise between the above conflicting requirements.

With respect to environmental effects, successive UK governments must surely have been aware of the health hazards from coal-fired power, but have been unwilling to publicise them for fear of admitting that a tremendous cost would ensue, even to achieve a small reduction. They have assumed that nuclear power would be justified on cost grounds and public acceptance of safety, so that there would be a steady replacement of coal-fired power and the ageing Magnox stations by PWRs. After Chernobyl, however, the implementation of this policy has become less straightforward. It is interesting to note that the UK has now begun to admit that UK stack discharges are responsible for Acid Rain (ITV, 1986); it will also be interesting to see in the future if sufficient admissions of other adverse effects of coal-fired power, such as the Greenhouse effect, are made just to ensure that the only real remedy - nuclear power - continues to replace it.

9.10 THE OPPOSITION

The Labour Party's statement on nuclear power, issued in September 1986, proposed a 'decades-long' phasing-out process, with a strategy based on coal, conservation and renewable sources. Magnox reactors were first priorities for closure, AGRs would be closed over decades to avoid power shortages, PWRs would be cancelled and Fast Reactor development stopped. No new contracts for reprocessing at Sellafield would be permitted; the THORP plant would be completed, but not commissioned for reprocessing. Instead, it would be used to develop the technology of waste disposal and storage.

The above proposals recognised the infeasibility of Greenpeace demands for the rapid closure of reactors as well as the consequent immense loss of sunk investments. However, if nuclear power were hazardous, the implementation of these proposals would not remove the risk, but merely offer a reduction over a long period. If then there was little urgency to remove the postulated hazard, why was it necessary for nuclear power to be phased out at all? It will be noted that no reference was made to the hazards of the proposed major substitute - coal-fired power. Once again, to politicians, jobs and votes were more important than deaths! If existing contracts, in particular the Japanese contract, were to be honoured, as might be presumed from the statement, then THORP would have to be commissioned for reprocessing; the latter would continue for a long period, under the terms of the contract. Subsequent modifications to refurbish it for developing methods of waste treatment and disposal would be extremely difficult and costly, if at all practicable. On the other hand, the penalties of reneging on the Japanese contract could be enormous and little of the plant, even if inactive, would be useful in waste treatment development. There is no mention of what the eventual fate of the stocks of plutonium and of the large accumulation of spent fuels would be, with the high radiological hazard potential of the considerable content of plutonium. Presumably, there would eventually be suitably smooth phrases saying why it had suddenly become acceptable to bury both the plutonium and highly active fission products. Meanwhile, the electricity consumer and/or the taxpayer would foot the bill!

Curiously enough, the dilemma of the Labour Party is the reverse of that of the Conservatives. Large units of nuclear power would naturally fall into the control of national bodies, traditionally favoured by Labour policies. Perhaps when the confusion between nuclear weapons and nuclear power has been dispelled, Labour may well look upon nuclear power kindly, particularly the Once-Through Cycle, which does not involve handling the 'dreaded element', plutonium!

The stance of the other Opposition parties is currently uncertain and subject to the reorganisation taking place between the Social

Democratic Party (SDP) and Liberal Party. However, prior to the 1987 election, the SDP sat firmly on the fence, by saying that it would not be prepared to sanction any new nuclear power stations until the Chernobyl disaster had been fully understood. Meanwhile, coal-fired power stations would be ordered. Research and development would continue on the CFR (presumably this means that reprocessing at Sellafield and/or Dounreay would continue in order to supply the fuel for the CFRs). There was felt to be no justification for Sizewell and PWRs in general. On the other hand, it was important to continue work on nuclear fusion, in particular the JET project. The above approach was suitably cautious and had the advantage of avoiding vote-losing attitudes. One would hope that the mists will have cleared a little before the next election. After all, in a politician's eyes, infinity is the interval between party manifestos and much can be forgotten meanwhile.

9.11 THE ECONOMY OF TRUTH AND THE 'S-J' CYCLE

The reader will readily recall the words of Sir Robert Armstrong at the 'Spycatcher' hearing in an Australian court, where he admitted that he had been 'economical with the truth'. It is a factor of modern life that tabloids and holoids try to exploit contentious issues to arouse the maximum sensation and controversy. Unfortunately, the teloids, though their constitutions specifically lay down that there should be due impartiality and a proper balance of information at all times (see Section 9.7.3), have followed a similar line to most holoids in trying to undermine confidence in nuclear power. Their persistent voicing of the views of their 'Green' consultants, together with a high profile of Greenpeace operations, can be contrasted with a silence on points in favour of nuclear power. Very briefly, they appear to have been thoroughly 'green-washed'. Overall, their attitude has amounted to a virtual censorship of pro-nuclear facts and views (see Appendix 9.3).

This persistent 'economy with the truth' by the BBC and IBA has naturally confused the average viewer, who has neither the time nor inclination to study the vast tomes of information presented by the Nuclear Club. The resulting concern generated in the viewing public is naturally passed on to their Members of Parliament who interpret it as 'public opinion'.

One of the problems of the media is that the public memory is short. For an issue covering a period of years, therefore, articles and programmes have to appear at regular intervals to try to justify earlier predictions and sustain what is then called 'increasing public concern'. In the case of nuclear power, sensations are few and far between. The anti-nuclear campaigns are therefore prolonged and most of the newspapers and the teloids become identified with an anti-nuclear stance which, if shown to be erroneous, would seriously reduce public confidence in their views. They in effect become locked in a sequence of Self-Interest and Justification, which, for convenience, we may dub the 'S-J Cycle'.

The reader will no doubt notice a similarity between Parkinson's Law, which states that 'work expands so as to fill the time available for its completion', and the S-J Cycle, in which the work of maximising financial returns is constantly rationalised. Both are related to the Pursuit of Progress (the alternative title of Parkinson's book) and satisfying the political stances of higher echelons. The cycle is not limited to anti-nuclear groups, however. Research groups in nuclear power face an ever-dwindling number of worthwhile problems. It becomes fatal to obtain satisfactory solutions when they may result in the reduction in numbers of research jobs (Appendix 9.1). It is, in fact, quite astonishing that the Nuclear Club proclaims that the complex plants it has developed and built (in only a few decades) are satisfactory, yet after ten to twenty years, large programmes of research are still needed into the vastly simpler technical problem of finding ways of disposing of nuclear waste. Perhaps some day, the question may be asked (by the taxpayer) as to the value of funding further research. Meanwhile, the public's problem of whether to believe the 'Number Games' of the Nuclear Club or the 'Word Play' of the 'Greens' and the media remains.

9.12 **OTHER COUNTRIES**

Although the main thrust of this book is concerned with UK power, it is important to observe developments in other countries to derive their impact back on UK plans. A useful background to record here is the percentage of national electricity produced by nuclear power in various countries in 1985 (corresponding forecasts for the turn of the century are given in brackets) - Argentina 11 (20), France 65 (83), India 2 (3), Japan 23 (39), South Korea 22 (22), Spain 24 (24), Sweden 42 (42), Switzerland 40 (65), UK 19 (25), US 16 (17), USSR 10 (13), West Germany 31 (31).

Those 'First World' countries with smaller natural resources of fossil fuels than the UK will wish to safeguard their energy supplies. This will entail constructing their own nuclear systems, since the advanced technology of their industries will require a reliable power supply at all times. The French, who have less fossil and water sources of power than most countries, have shown great enthusiasm for nuclear systems. In France, PWR fuel is manufactured by a Franco-Belgian consortium (FBFC). A second Franco-Belgian company (Commox) has been set up to prepare oxide fuels with 3-5% of plutonium in uranium. Reprocessing of PWR fuel is carried out by Cogema at Cap de la Hague in Normandy. Overseas countries sending spent fuel for reprocessing in France include West Germany, Belgium, Holland, Japan and Sweden. The waste disposal of French LLW and ILW is run by ANDRA, the French equivalent of NIREX. HAL is vitrified for storage at Marcoule.

Not surprisingly, because of the shortage of indigenous fossil fuels, in addition to existing reactor systems, Thermal Reactor fuel reprocessing plants are also planned for West Germany and Japan. Further, if a Fast Reactor system ever becomes attractive to the UK, it will surely also be attractive to the countries mentioned above. The desirability of a prompt turnround of plutonium for Fast Reactor fuels within about a year, including several months cooling before transport is possible, means that the same countries would build reprocessing plants reasonably close to the reactors they serve.

In the US, the availability of cheap fossil fuel lessens the driving force towards nuclear power. Though many reactors have been built and are under construction, the protracted planning negotiations and complicated regulatory procedures have slowed down the substitution of coal by nuclear power. Another important factor here is the US legal system. Litigants do not necessarily pay the costs of both sides when they lose. Further, contingency fees may be accepted by lawyers, i.e. a percentage of the winnings and no fees on losing. The claimant is therefore at little financial risk; because of this, companies may buy off such vexation, or in the analogous case of (say) a new drug, not risk putting it on the market (Section 9.8).

In Sweden, a curious dilemma has developed. Although it has been decided that nuclear power is to be phased out, the admitted loss of cheap power will hit papermill, steel and aluminium industries: yet the environmental lobby is so strong that more coal power is unlikely and there is also opposition to the development of further hydropower sites.

In developing countries, the high technology and concentration of power in large units, demanding strongly connected and reliable transmission systems, is for some regions less appropriate than energy from traditional sources such as wood in small combustion units. In many cases, however, such fuels are being used up much faster than they can be replaced by growth. The average citizen of the Third World consumes fossil fuels at only a fortieth the rate of the average person in the UK. With increasing world population and standards of living, the world demand for energy could therefore multiply severalfold in the next fifty years. Realistically, in the poorer countries, even with a big expansion in the use of renewables and a considerable increase in the efficiency of energy use, much of this increase will have to come from oil.

Currently, therefore, the use of renewable energy processes (which could include solar power in 'high sunshine' regions relative to the UK) should expand in developing countries. This is particularly true for wind power, though the present rate of breakdown of wind machines is too high for wide application. In many areas of developing countries, the unsophisticated lifestyle could well find the intermittent power supply characteristics of renewable processes acceptable. Conversely, reduced energy requirements could arise if the present highly inefficient methods of cooking were improved through the use of stoves of better design.

The possibility of radical regimes gaining control of a country is the 'unacceptable face' of nuclear power proliferation. One wonders what the present government in Iran would have done with the

18 nuclear reactors scheduled to be built by the French for the Shah's government, if the revolution had happened 10 or more years later. On the other hand, in public debates between pro- and anti-power groups (Foley, 1978), neither side appears to make out a credible case that either UK nuclear expansion or contraction would alter corresponding power systems in the rest of the world.

9.13 SUMMARY

In the early stages of nuclear power, large and diverse organisations were set up for development and construction. The driving force was associated with the desire to break the dominance of coal in power supply. With the need for co-ordination and large groups to face up to overseas competition, amalgamation resulted in one reactor construction company, NNC; BNFL was the sole British fuel cycle company, so that eventually there was little competition to these two organisations in the UK. In recent years, the discovery of further low-cost uranium ores has reduced the urgency of developing Fast Reactor systems. With little necessity too for new Thermal Reactor designs, the workload of the UKAEA has fallen appreciably and much emphasis has been directed towards waste treatment development for BNFL and waste disposal work in support of NIREX. The case for recovering plutonium from spent Thermal Reactor fuels to charge into Fast Reactors having diminished, reprocessing is thus less urgent. BNFL may now have to base its case for continued reprocessing through, on the one hand, the reduction of eventual waste disposal problems by recycling plutonium in Thermal Reactors and, on the other hand, the difficulties of stopping reprocessing, storing spent fuel for long periods and subsequently building up the capability to restart operations.

With the reduction in importance of the NUM after its defeat in the 1984-5 strike, the fear of its domination of the coal industry has receded; with it some of the impetus to develop nuclear power further has faded.

Of the developing systems, fusion requires very large experimental units for testing possible power processes, so that considerable international co-operation and long timescales for research are necessary. Several companies are offering commercial wind machines in the 60 to 300 kilowatt range; larger machines are under test in the UK, but there seems to be no concept of energy storage to overcome the shortcoming of intermittent power supply. 'Wind farms' are planned in various parts of the UK, but large areas are needed, many in regions of designated natural beauty. Wave systems, particularly those located on the coast, are under active development but will probably be appreciably more expensive than wind systems; some Norwegian concepts have interesting features.

In the electricity supply industry, the UK supergrid is centrally controlled by the CEGB with area distribution under twelve Boards. The CEGB carries out research in aspects of various power systems, in particular (with British Coal) on developing cleaner stack discharges from coal-fired power stations. These organisational arrangements will probably be altered considerably if Government privatisation proposals are carried through.

In contrast to the large outputs of data and numerical assessments by the Nuclear Club, the 'Green' organisations, FOE and Greenpeace, base their attacks on nuclear power on descriptive criticism. There is little numerical content on which the respective points are founded; nor is there any comparison between complete power systems. The media has discovered that, regardless of the relative merits of the various power systems, nuclear scares make good headlines and many articles exploiting this have maintained a distorted image of nuclear power in the minds of the general public. In spite of the widely differing views of the main political parties on the future role of nuclear power and the clear requirement in their constitutions to provide balanced views with no political bias, the BBC and IBA have carried out a sustained campaign highlighting what they consider to be adverse features of nuclear power. It is then claimed that the public is 'getting concerned' about nuclear power, i.e. the campaign is successful. In contrast, there are no programmes on the high rate of accidents and diseases to miners or the health detriments to the public of coal-fired stack discharges, except through secondary effects of Acid Rain.

The length of the anti-nuclear campaign in the media has brought about its own problems. Though 'doom-laden' forecasts persistently remain unfulfilled, it is becoming increasingly difficult to withdraw from the 'S-J' Cycle, without the admission that the campaign was wrong, thus losing considerable public credibility.

The analysis of the response of the general public to the chance and consequence of low-frequency events has expanded in recent years into a new science of 'Public Perception of Risks'. There is a natural desire from the public for sensation, satisfied by the media for an appropriate fee. The exaggerations by the media in turn instil in the public an impression of frequency of rare events much higher than is actually the case. Unfortunately, such imagined effects can have real impacts on (say) the value of property or produce near an area designated for nuclear waste disposal.

It is perhaps useful here to recapitulate some features of the history of UK nuclear power. Successive Governments, democratically chosen by the electorate, have backed the development of nuclear power. A considerable number of carefully chosen recruits, with appropriate academic training funded by the taxpayer, have gained long-term skills to build and operate nuclear plants to provide a lower power cost than that from coal. If such approved teams, with separate watchdog experts such as NII, now recommend nuclear power as better in cost and safety than power from coal, to whom should the public listen - to these or to the views of self-appointed Green pressure groups of uncertain qualifications, backed by the media? If the views of the latter are preferred, what is the point of electing Governments, or indeed of education itself?

Successive Governments attempt to appear to be 'doing something' by 'doing research' on nuclear waste disposal, but little seems to come out of this but more research. The time for awkward political decisions is always after the next election! The 'softly, softly, catchee monkey' approach of replacing coal by nuclear power has suffered a rude shock from Chernobyl and the media hype that followed. Some 'knocking' of coal power is now to be expected with costly cleanup systems at last being installed in the UK to reduce the pollution outcry from the rest of Europe. On the other hand, the present Opposition parties have put out 'cosmetic' policy statements on running down nuclear power slowly, which itself is an admission that there is in reality little hazard.

Overall, the various parties involved in UK power tend to work out an 'S-J' Cycle of first establishing where their self-interest lies and then later attempting to justify it. Though this cycle is particularly pronounced in the 'Greens' and the media, there are corresponding characteristics within the Nuclear Club.

Outside the UK, 'First World' countries with little natural resources for fossil and hydro-power have no option but to develop nuclear systems relatively rapidly to assure a reliable supply of power. For example, France is generating three-quarters of its electricity by nuclear power and the installed power in Japan will be roughly doubled in the next decade. For developing countries, the large centralised blocks of electricity generation characteristic of nuclear power seem inappropriate in general. Renewable energy supplies, though intermittent, could be acceptable to the less advanced industries and with more efficient methods of energy use are probably the best option to conserve fossil fuel supplies. It is arguably in the interests of technically advanced countries, to whose lifestyle the reliable and environmentally clean nuclear power is appropriate, to assist regions with less complex lifestyles to install renewable power systems rather than fossil fuel burning. This would help to conserve forests and not to exacerbate the world emission of carbon dioxide and its associated Greenhouse effect.

In the long-term global future, supplies of energy are far greater from Fast Reactor systems than from Thermal Reactors, coal, oil and gas: renewable resources are, of course, potentially limitless, but cannot provide sophisticated industries with a reliable continuous power supply.

Appendix 9.1

NOT INVENTED HERE (OR RESEARCH WITHOUT END)

After the early days of big expansion of nuclear programmes, with the corresponding buildup of large development and construction organisations, the subsequent slowing-down led to a scramble for the correspondingly reduced funds. The original palmy days of rapid staff expansion were replaced by a saga of redundancy and empires on which 'the sun never seemed to stop setting'. The urgency to resolve problems correspondingly diminished. Nuclear waste, in particular, became an area where decisions generally caused more political trouble than did prevarication. The time for finding solutions to its problems therefore expanded nicely and work was smoothly generated in the Parkinsonian mode and channelled (predominantly) between Government organisations to fill all the time and funds available. Neither opponents of nuclear power, nor the media, nor research groups were keen to obtain the final solution, as controversy engendered public interest and money. It became important in the scrabble for work to convey an air of superiority, ignoring suggestions from other groups with the well-known 'Not Invented Here' or 'NIH' syndrome. In this respect, a frequent complaint from the 'Greens' was the 'Nanny Knows Best' attitude of the Nuclear Club. This was hardly surprising since, if Nanny did not appear to know best, the taxpayer might wonder why Nanny was being paid! In the words of Parkinson (1958, p.95), there arose the perfect ingredients for the onset of injelititis, or palsied paralysis.

An example of the above was the development of highly sophisticated computer programs 'crunching ever larger numbers' in the prediction of the behaviour of disposed nuclear waste beyond the next Ice Age. This was in spite of the lack of firm data on rock characteristics or in some cases, the composition of the waste.

The following anecdote seems relevant to nuclear waste research. During a visit to a research laboratory of a large industrial organisation, a chemist was asked how the new project was going. 'Slowly,' was the reply. 'This problem is very interesting, so I shall certainly be in no hurry to solve it!'

Appendix 9.2

SPRIGS OF GREENERY

A common approach of the 'Greens', in their anti-nuclear stance, is to string together a few unconnected emotional phrases, e.g. the 'plutonium economy', 'the most radioactive sea in the world' (i.e. off Sellafield) and repeat these through numerous newspaper articles and television interviews, backed up by stunts carefully prepared for notice by the media. A report in 1986 by the House of Commons Select Committee, however, with members of a wide variety of political outlooks, showed no illusions about 'the tactics resorted to by some pressure groups in order to attract attention to their cause.'

An example of misleading exaggeration is a full-page Greenpeace advertisement in the Guardian, cataloguing trivial incidents at Sellafield; strangely enough, no one appears to have been seriously injured or subjected to a significant health hazard. How odd that they did not compare this list with those of a similar workforce in other industries, in particular the number of deaths and serious injuries in the coal industry equivalent to the same output of power! Another example of a complaint 'in isolation', i.e. without reference to similar operations, concerns protests about May and Baker discharges of 24 kilogrammes of mercury per year into Norwich sewage (Greenpeace, 1986), whereas the current annual output via coal-fired power stacks into the UK atmosphere is 100-200 tonnes (Table 3.2), a factor of several thousand greater.

The original 'Green' outcry against leaving nuclear wastes to be dealt with by future generations has now been joined by a plea for long-term storage of spent fuel (Chudleigh, 1984). One wonders if the latter should be labelled 'Best Reprocess After AD 2500'.

More recently, an example of the attempts to 'knock' nuclear power by trying to rake up old history is the article by Patterson (1988) in the Guardian. The title, 'Lies, Damned Lies and Magnox' is particularly strident; the content, however, is 'old hat' on the valuation of plutonium in the 1960s and the possible use of civil Magnox reactors to make military plutonium.

Appendix 9.3

THE DEVELOPMENT OF THE BRITISH MEDIA

An excellent account of the development of the Press and Broadcasting in the UK is given in Curran (1988). This book states,

> the view that the British press is one of the great instruments of liberty ... the vital defender of public interests, is a central part of our political culture ... The theory was produced to justify those who created the press and whose interests it largely served. This does not mean that newspapers, television and radio have generally been instruments of crude propaganda; rather that the media are political actors in their own right ... Arguably, the power of the media has increased remarkably in the last forty years. There are less alternative sources of information, while the control of the media has become concentrated in fewer hands ... It has become less accountable ... Thus the press and broadcasting exercise a massive power, but it is more than ever a power without responsibility.

The Third Royal Commission on the Press in 1977 stated bluntly, 'anyone is free to start a daily national newspaper, but few can afford even to contemplate the prospect.' As a result, there has been a concentration of ownership which has contributed to a much greater convergence of opinion within the Press. In this respect, Curran instances the unanimous national newspaper attitudes in supporting trade union 'reform' in the 1960s, the urging of Britain's entry into the Common Market in the 1975 Referendum, the backing of Jim Callaghan for the Labour leadership in 1976 and the opposition to the TUC's 'day of action' in 1980. Similarly, an analysis in 1977 showed little difference in the reporting of industrial relations between left- and right-wing papers. Management was relatively invisible in Press coverage - only the trade unions were called upon to account for their actions. Effectively, strikers were portrayed as being in conflict with the public rather than with their employers.

As costs rose in the 1950s and 1960s, mass marketing pressure steadily reduced political coverage in the Left and popular Press. By 1976, none of the seven popular papers devoted more than 20% of their editorial comment to public affairs. On the other hand, the quality Press continued to maintain a high level of political coverage. However, a research survey over the period 1963-71 showed that the most read items in all types of daily and Sunday newspapers were human interest stories. Increasing circulation by catering for such popular tastes may not be an automatic objective, since much of the revenue, particularly for quality papers, comes from advertising. Indeed, in the late 1960s, the Times adopted a deliberate policy of shedding 'popular' readers and directing their efforts to increase sales to the type of reader which might be more preferable to advertisers. As a consequence, sales fell from 430,000 to 340,000 in two years, but profitability increased.

There have been many analyses of the degree of influence of the media on public opinion. It has been found that this influence is related to the trust in the source of the message, particularly in subjects where the public has little direct experience.

In an effort to reform the Press from within, the First Royal Commission on the Press proposed the setting up of a Press Council. This was reluctantly established in 1953, but only after threat of legislation: members were restricted to Press representatives. Its conduct was sharply criticised by the Second Press Commission after which the Council enlisted a larger lay membership. The Third Commission was still strongly critical, hoping that in future the Press would be 'more vigilant in demonstrating the independence and impartiality to which it lays claim.' The Commission made twelve recommendations for the reform of the Press Council: nine of these were rejected, including key proposals such as securing front page or 'equal prominence' publicity for complaints upheld by Council with respect to offending publications. Few changes were made in line with the other recommendations. Indeed, in 1980, the National Union of Journalists withdrew from the Press Council because the latter was neither effective nor genuinely independent of Press management.

On British broadcasting, there was an assumption of commitment to an undivided public good in official thinking on radio and television until the 1970s, i.e. presenting a middle ground of opinion on which most people could agree. However, in 1977, the Annan Report replaced this with a competing multiplicity of independent voices. The result has been confusion and crisis (Curran, 1988).

In 1960, the Pilkington Report argued that the problem of how to make broadcasting politically accountable, yet free of political influence, particularly by the Government of the day, was made easier because the first priority was to resolve technical matters. No conflict had arisen between broadcasters and Government over the definition of public interest. By the 1970s, however, the relationship between the State and broadcasting had become steadily more hostile. This arose partly because of a proliferation of parties, interests and pressure groups. The situation in Northern Ireland is an example where 'balance' in dealing with treasonable activities and the associated dilemma of whether to interview terrorists, forced broadcasters to make their own rules. Indeed, by 1977 the IBA claimed that accountability was only a minority interest.

Curran concludes that broadcasters have come to see the State as their enemy. Consensus on what is the 'middle ground' of agreed opinion has broken down. Broadcasting needs to find a new relationship with the State and a new form of commitment to public service. Among proposals for reforming the newspapers, it is suggested that the Press Council should be more representative of Press employees and the general public. Where the Council upholds complaints, these should receive at least equal prominence to the offending article. An analogous Broadcasting Council, operating as mooted above for a reformed Press Council, should be set up.

Appendix 9.4

MEDIA CENSORSHIP

In the last few years, the anti-nuclear stance adopted by the media has become more and more pronounced. The slightest incident at Sellafield has been highlighted as an important news item, in spite of the fact that no operators have been killed and the general health record bears favourable comparison with that of any similarly sized industrial organisation. On the other hand, though hundreds of new lung disease cases from coal mining are notified each year and several deaths from both this cause and mine accidents occur every month, these never reach the media headlines (see

Fig. 9.2). A similar 'man bites dog' attitude holds for hazards to the public: tremendous coverage has attended events at Three Mile Island and Chernobyl, but virtually nothing has ever been said about the health effects of routine discharges from coal-fired power stations.

Some examples of media distortion follow. A Sunday Times colour supplement article in 1986 was entitled, 'The Radioactive Sea', by Geoffrey Lean and Walt Patterson. It is claimed that 'the sea that laps this (Seascale) beach contains a quarter of a ton of plutonium'. Later, it says, 'this point has not been lost on the Irish Government' and 'plutonium from Windscale has been identified as far away as the North Cape of Norway and the waters off Greenland'. A diagram, suitably with red arrows, shows the main flow of activity passing between *northern* Ireland and southern Scotland on its way to Greenland and the Norwegian Sea. It is obvious nonsense that the sea near Seascale contains 'a quarter of a ton' of plutonium, since the currents will sweep discharges to the north as shown in the diagram. Only a small fraction is held back, not in the sea, but in sediments on the sea-bed and shore. Even these in fact exchange plutonium back to the sea to be carried away north (Section 6.2.4). The 'quarter of a ton', which is the total plutonium from many years of very dilute discharges, should hardly concern the Irish, who are sitting comfortably upstream of the discharges! The article continues along the well-worn trail of linking together all the 'mishaps' of nuclear power without stating the consequences: spent fuel being unloaded at Barrow is, as usual, described as 'imported nuclear waste'. The mention of detection of plutonium in Arctic waters fails to note that the instrumentation for measuring plutonium is so sensitive that the limits of detection are many powers of ten below any hazardous concentration in seawater. One wonders what McLean and Patterson believe of the fate of untreated Seascale sewage which discharges tens of thousands of tonnes of human excrement closer to the shore than the end of the Sellafield pipeline. Is swimming off Seascale merely going through the motions?

Other articles in the Press follow similar lines. David Fairhall, in 'Why You Can't Get Away From It All' (Guardian, 6 June 1986), gives radiation zones of 100 miles round 17 nuclear stations in the UK to illustrate the consequences of any of them becoming a 'UK Chernobyl'. In fact, since there is only one accident under discussion, though admittedly very serious, the affected zone occupies only a fraction of the UK. Unfortunately, coal-fired power station stacks are well distributed around the country and built very tall to ensure the effluents are uniformly spread. Perhaps some day David Fairhall will write another article on coal power explaining why you can't get away from it *at all*!

A further example in November 1986 (Tomorrow's World, 1986) is a discussion on how the UK might 'do without nuclear power'. This is a 'leading' title; a fairer subject would have been 'What is the best combination of power systems for the UK in the future?' The programme was virtually a restatement of the so-called strategies put out by the BBC's consultants for the programme, who were Greenpeace and FOE devotees. Eight estuaries were suggested to provide tidal power, and with many wind turbines (ten of them on the Isle of Man alone), these would provide 10% of UK future power. No mention was made of disadvantages such as the noise and unsightliness of wind machines, loss of birdlife in tidal schemes, the level of mismatch of CHP heat and power demands or the fate of British industry during climatic phases when no power was available from the renewables. The pollution from the bulk of the power - supplied by coal-fired stations - was, of course, not discussed.

Another example of a biased programme was 'The Dump' (BBC2, 1986), which showed the reaction of the people of Fulbeck in Lincolnshire to NIREX tests to ascertain if a local disused airfield was technically suitable for a national LLW disposal site. Note the title, which is much more appropriate for the spoil heaps from coal power than the careful emplacement of securely packaged wastes into concrete cells. Strange, too, that no interviews were included with the inhabitants near Drigg in Cumbria, who have had the national LLW disposal site with simple burial in soil on their doorstep for over 30 years!

A common attack on nuclear power is to hark back to the early days of Magnox reactor operation (David Taylor, Taming The Dragon, BBC2, 8, 15 and 22 October 1987). Here, claims are made that civil Magnox reactors were designed to be able to make 'military quality' plutonium. Since this only meant discharging fuel after a short irradiation, it would seem difficult to avoid such flexibility, which was probably necessary anyway to discharge the occasional faulty fuel element. Whatever the true facts on whether and to what degree civil reactors were used for making 'military' plutonium, there was at that time a national requirement to produce such plutonium and the forecasting of military needs would be uncertain. It would therefore have been imprudent to have excluded by design the possibility of making 'military' plutonium in civil reactors, at the same time building spare 'military' reactors just in case more plutonium was needed! Taylor's theme has been reiterated many times in the media, e.g. Patterson (1988).

Unfortunately, the IBA follows the BBC in striving to link atomic bombs with nuclear power. (An apt comment by Franklin in Foley (1978) on such a linkup is that, 'no one stops using wheels just because they were first used on chariots'.) In a recent Channel 4 'Under Fire' programme, the former Chairman of the UKAEA and BNFL, Sir John Hill, was carefully introduced as having a part in the development of such bombs, whereas the Director of FOE, Jonathan Porritt, was introduced as a 'double first' i.e. a very clever fellow. Since the subject was nuclear power, the mention of bombs was irrelevant. Moreover, it was not certain whether Mr Porritt's qualifications were in appropriate subjects, or indeed significant in relation to the thousands of highly qualified and experienced staff who had been under Sir John's control.

A further example of the anti-nuclear stance of companies under the control of the IBA was the programme by Yorkshire Television, which first highlighted the leukaemia 'cluster' round Seascale. The number of excess cases over a period of several years was only four, but the programme was described by its producer, James Cutler, as part of a crusade. It is unclear whether the gallant Mr Cutler mounted his white charger before he established any 'facts' or whether he later accepted the findings of the experts in the ensuing Black Inquiry; what is strange is that, in view of the hundreds of deaths caused annually from coal-fired power stacks, Mr Cutler and a hundred 'Green' knights are not to be seen charging up these stacks to pop a plug in the top of each! It would seem a far more efficient use of their 'anti-pollution' energies than blocking the Sellafield pipeline. Recently, doubt has been cast on previous statistical methods of establishing the existence of leukaemia clusters (Openshaw, 1988 and Taylor,

1988). Openshaw (1988) describes a new procedure using detailed analysis on large computers which reveals that, in the North of England, several clusters exist (see Appendix 6.1). In one of them, at Tyneside, there were 165 cases in 18 years - far greater than at Seascale. There are no nuclear facilities within 35 miles of Tyneside and Openshaw concludes, 'it is likely that the causes of the leukaemias are perhaps pollution-induced and not related to radiation.' In line with the media's censorship of favourable features of nuclear power, there was no rush to broadcast this news, unlike the scramble to highlight any article apparently linking nuclear installations with leukaemia. Similar front-page and prime-time coverage (Fig. 9.3) was given to the COMARE report on Dounreay described in Appendix 6.1 (typically a 46 square inches leading front-page article in the Guardian) but no comment on the critical review by Taylor (1988). A later programme by Cutler in 1988, attempting to relate leukaemia clusters to radioactive discharge from non-nuclear industrial plant, was found to have no sound basis, as recounted in Appendix 6.1. The IBA did not see fit to give a follow-up programme with the independent assessor's findings. The Times, to its credit, announced the findings in an article of 16 square inches at the bottom of page 2, 12 July 1988; other holoids published nothing. It may have escaped the moguls of the media that people's lives are at stake, not just the jobs of journalists. An appropriate campaign would have been to demand more support for Dr Openshaw and his colleagues to identify the cause of the leukaemias by his detached, scientific approach.

Appendix 9.5

THE IMPROVEMENT OF PUBLIC UNDERSTANDING

There is, of course, a wide gulf between what is technically desirable and the acceptance of such by the public, who may continue to receive a largely distorted picture from the media. Ideally, it would be advisable to introduce 'seatbelts' for journalists and presenters of programmes: the need for freedom in the media should not be used to justify the doctoring of the truth. One form of this could be for an Ombudsman or some form of regulatory group (say a cross-party parliamentary panel) to be empowered to insist on a corrected version of an article or programme being issued in exactly the same way as the offending one. For example, in a newspaper, the same area and position on the same page would have to be used, whereas for television the same length of programme at the same time and day would have to be broadcast (see also Curan's observations at the end of Appendix 9.3 and the proposed setting up of a Broadcasting Standards Council in Section 9.7.3.)

The gathering of authentic data for such public presentations should not necessarily be left to the media, since the interpretation of the results is prone to both bias and genuine error. The present system of confrontation between parties of opposing views, egged on by a media chairman, should be replaced by a detached and learned review of the issues. It should not be too difficult to set up a body of unbiased expertise, perhaps mainly from universities. (After all, what are seats of learning for, if not to educate?) The resources of the media should then be instructed to back up in the presentation, but final decisions on content would rest with the panel.

One topic worthy of detailed study is the variation of health effects with distance and direction (especially relative to prevailing winds) from coal-fired power stations. Both respiratory diseases and cancers should be included: direct comparisons with (say) the hazards of under-sea-bed plutonium and neptunium are important. The work of Fremlin (1985) is, of course, a good indication of what can be done, but the subject deserves considerably greater effort. A comprehensive study of depositions from stack discharges (possibly using their radioactive components as tracers) should assess the pathways back to man through inhalation, grazing, livestock, etc. These results could then usefully be fed into public enquiries on construction of new coal-fired stations (including the dumping of associated wastes).

Meanwhile, there is a series of demonstrations or 'counterstunts' which might usefully be carried out to increase public understanding and confidence in the management of nuclear waste, in particular to show that any delays in implementing disposal have been for sound technical reasons. (West Germany, India, Sweden and Switzerland have firm timetables for such demonstrations.) The first of these relates to verifying that future generations will have at least one practical method of disposing spent fuel if the Once-Through Cycle is used. Other demonstrations relate mainly to aspects of nuclear cycles involving reprocessing.

(a) Deep holes should be drilled and dummy fuel elements (containerised as necessary) lowered down them. Simulated accidents and associated remedial operations should be carried out. Typical locations could be at Dounreay and Sellafield with boreholes slanting under the sea-bed to a depth of 2 to 3 kilometres. It would be important to demonstrate that the disposal zone was well on the seaward side of the saline/freshwater interface in each case (see Appendix 4.3). The dummy element tests should then be followed by runs with real spent fuel elements (AGR or PWR, not Magnox) with appropriate monitoring of temperatures, activity movements, etc. Similar tests could be done with HEW glass blocks. The experiments should, of course, be designed to fit in with a follow-up demonstration programme of permanent disposal if all went well.

(b) Tunnels typical of the Dry Repository should be excavated on preferred disposal sites (Section 4.1.8). Spent fuel, glass blocks and cemented packages could be tested in these. A stagewise demonstration could include the checking of water flows into empty tunnels, the effects of dummy units containing heaters, followed by realistic tests with radioactive packages.

(c) The decay of radioactive isotopes in vitrified and cemented wastes can be quite accurately estimated. Consequently, for example, compositions of 100-year and 1,000-year glass can be predicted and the corresponding glass blocks manufactured. Additionally, external irradiation of glass blocks could simulate decay energy damage over a long period. Leach water from these blocks could then be checked for potability - a convincing public relations exercise would then be for senior nuclear executives to be seen on television sitting on the shielded blocks and drinking the leachates.

(d) Corresponding tests could be done with glass blocks of various 'ages', e.g. 100 years and 1,000 years, in seawater; fish could be reared in seawater circulated over the blocks and then monitored for human consumption (again preferably tested by senior nuclear executives).

(e) Long-term simulations could be done with seawater 'doped' with neptunium and plutonium (as if it had escaped from the sea-bed). Shaking with typical sediments and/or groundup rock from a probable disposal point could establish pickup properties; again, various types of fish and molluscs could be monitored in such solutions and eventually consumed.

(f) An offer could be made at an early date to dispose of some radioactive waste to the same standard as waste on coal dumps, i.e. dilution of the wastes in (say) soil and then spreading the latter over the ground in the manner of a coal spoil heap, to be leached into groundwater and local water courses.

Figure 9.1: MAJOR POWER STATIONS IN ENGLAND AND WALES AND SOURCES OF FUEL

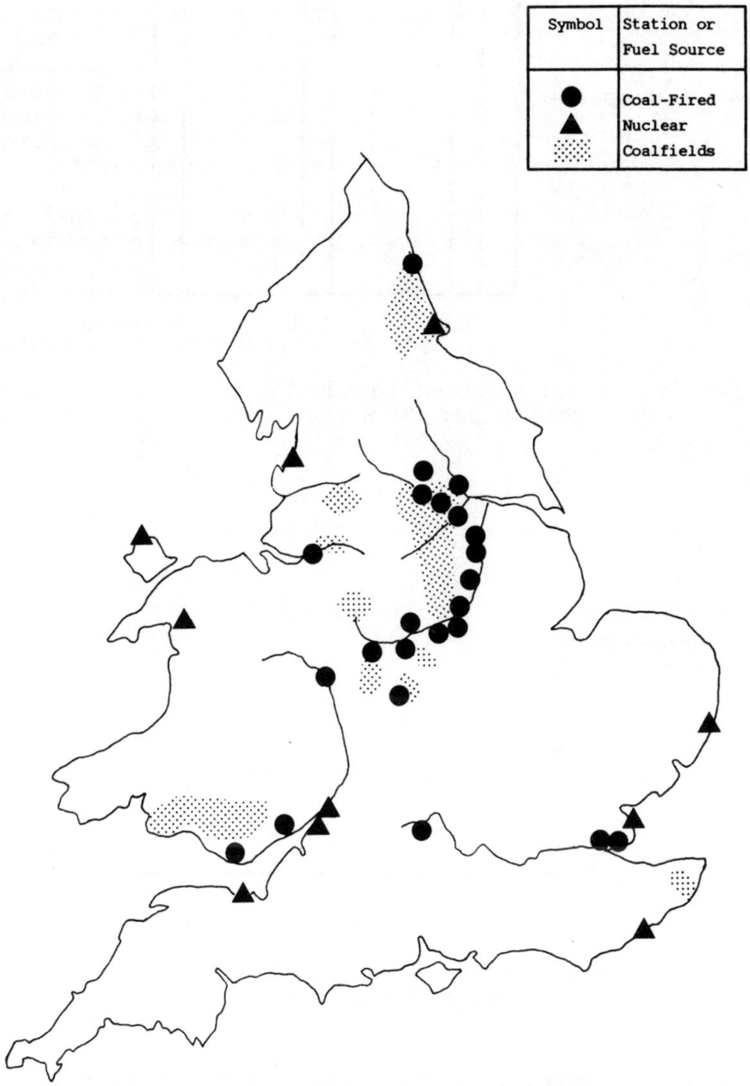

Figure 9.2: 'ECONOMY OF THE TRUTH' IN HOLOID AND TELOID REPORTING OF POWER OPERATOR DEATHS

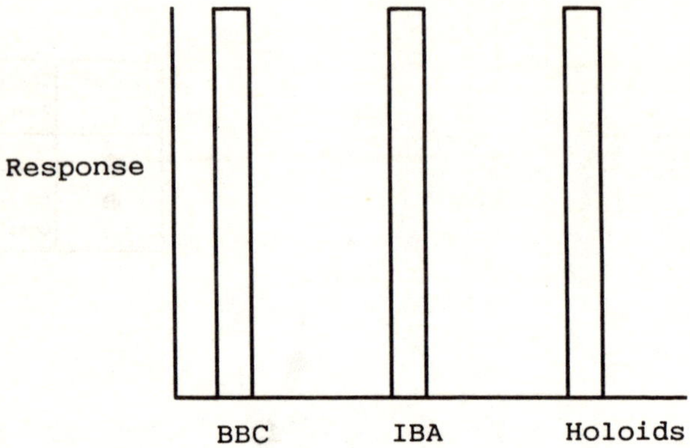

(a) Sellafield Nuclear Incidents
(<0.1 Deaths per GW(E)yr)

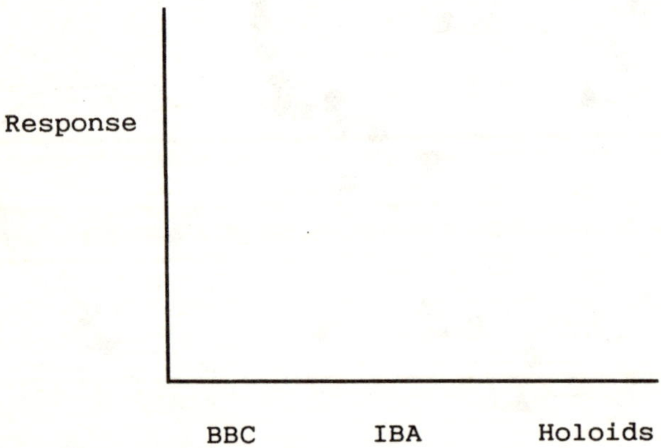

(b) Miners Killed in Power Production
(1 Death per GW(E)yr)

Notes: The response in (a) indicates instant coverage on TV main news bulletins or a prominent front-page holoid article.
The response in (b) indicates no mention on TV or in holoid articles.

Figure 9.3 'ECONOMY OF THE TRUTH' IN HOLOID AND TELOID REPORTING OF LEUKAEMIA CLUSTERS

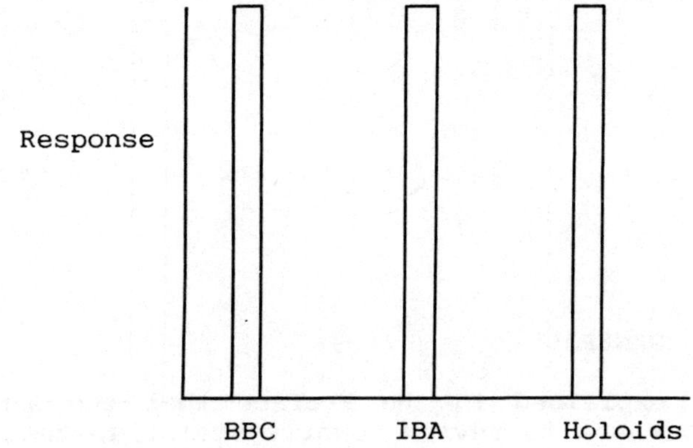

(a) Leukaemia Cluster Near Nuclear Site
(6 cases near Dounreay)

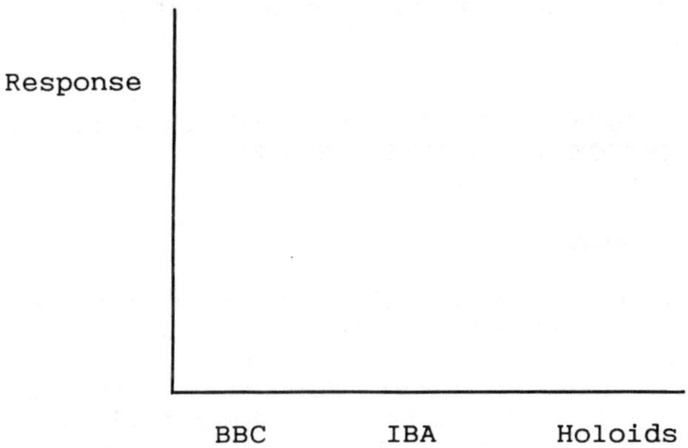

(b) Leukaemia Cluster Far From Nuclear Site
(165 cases on Tyneside)

Notes: The response in (a) indicates a principal item on TV main news bulletins or a prominent front-page holoid article. The response in (b) indicates no mention on TV or in holoid articles.

Chapter Ten

THE WAY AHEAD

10.1 GENERAL

It was explained in the Preface that the aim of this book was to review power systems, so that the reader could form a judgement on possible ways forward for the UK. Earlier chapters have described technical, financial and political factors associated with power systems. We recapitulate on these features in the following section, from which conclusions are then drawn in Section 10.3. A possible strategy for UK power is derived from the conclusions and set out in the final section.

10.2 A SUMMARY OF FACTORS INFLUENCING THE CHOICE OF UK POWER SYSTEMS

10.2.1 General

In the sections below, the origins and properties of various possible systems for UK power are presented in comparative form, followed by financial and political aspects. As far as possible, this is done diagrammatically. The ranges over which the relative properties vary are considerable and would, in normal scientific papers, usually be set out on logarithmic scales. However, as some of the relevance would then be lost, linear scales are used which show up the differences between systems more clearly.

10.2.2 Sources of Power

In a supernova explosion billions of years ago, a gigantic series of nuclear reactions took place, resulting in the creation of a large range of nuclides. Some of these came together with primordial matter to form the Earth, some five billion years ago. Many of the nuclides were unstable and gave out radioactivity as they decayed. An important example of this is plutonium-239, which has a half-life of 24,400 years; a negligible proportion of its original nuclei therefore remains to the present day. However, a considerable fraction of some radionuclides with very long half-lives - thorium-232 (14 billion years), uranium-235 (0.7 billion years) and uranium-238 (4.5 billion years) - has survived, continuously generating sets of other radionuclides (see, for example, the uranium-238 decay chain shown in Fig. 1.1). This natural radioactivity yields a continuing output of heat in the Earth's crust and so is partly responsible for the availability of geothermal energy and conditions suitable for the formation of coal and oil.

Our Sun is a 'medium-sized' star which generates much of its energy by fusing together hydrogen nuclei to create helium nuclei. Similar reactions are currently under investigation to develop fusion power on Earth. The fuel supply for this looks almost unlimited, but the technical problems are so severe that large operating units providing continuous power on demand seem many decades away. The intense heat created by the above fusion reactions in the Sun's core raises its outer surface temperature sufficiently for an enormous quantity of visible and other radiation to be emitted into space. Some of this radiation can be collected in the UK and used as solar heat. The Sun's radiation also evaporates large volumes of water, returned to the ground mainly as rain. Storage of the runoff in mountainous regions in the UK can be used to operate turbines in its fall to lower levels, i.e. to create relatively cheap hydro-electric power. Unfortunately, there are few regions left in the UK where further exploitation of this power source is worthwhile.

A further effect of the Sun is its gravitational pull on the Earth, keeping it in orbit and drawing up the oceans. A greater effect of the latter kind is caused by the Moon, which brings about the twice-daily rise and fall of the tides; particularly high tides occur when the Sun and Moon pull in line. This cycle of changing heights of the surface of the sea can be controlled by a barrage to operate turbines and produce tidal power. Such a system is feasible in several locations in the UK, especially in the Severn Estuary.

The rotation of the Earth and the Sun's radiation create currents in the atmosphere; these winds can be made to turn shafts and so generate power. Winds can also have an effect on the surface of the sea, creating waves; the latter can then be used to operate turbines for power. Like geothermal, solar and tidal energy, 'free fuel' is available indefinitely, but both wind and wave power can only operate in the right climatic conditions. Unfortunately, too, large energy storage schemes to bridge the 'calm periods' are generally impractical in the UK. An interesting consequence of the need to achieve favourable climatic conditions is that, although individual wind and wave units are relatively low in output, they will often be grouped. Effectively, therefore, they provide blocks of power.

Some of the nuclides still in existence on Earth can be used to supply power by being fissioned with neutrons in controlled chain reactions. The main system today, using slowed-down neutrons in Thermal Reactors, depends on the isotope uranium-235, present naturally at about 0.7% of uranium in ores. In the early days of nuclear power, the supplies of low-cost uranium ore appeared limited with respect to possible world demand, indicating that a large expansion of Thermal Reactor power would entail the mining of ores which were difficult and expensive to recover. By extracting plutonium from spent Thermal Reactor fuels, however, a reactor using neutrons at high velocities can both fission plutonium and create from uranium-238 sufficient new plutonium for further reactor fuel charges. The overall effect of such Fast Reactor operation is to consume the common isotope of uranium (mass 238), thus offering the prospects of sixty times more energy than Thermal Reactors from the same amount of uranium. Abundant energy for the world is thereby potentially available from Fast Reactor systems. In recent years, however, the discovery of large supplies of low-cost uranium ores suggests that Thermal Reactors could be run economically for a long time, thus reducing the urgency to bring Fast Reactors on line.

Vegetation laid down millions of years ago has been changed by heat and pressure to coal, familiar as a major source of heat in the more developed countries for many decades. The chemical energy converted to heat by the combustion of the coal is far less per atom than that of a nuclear fission reaction. For a given energy requirement, therefore, the physical side of all operations of coal-fired power is very much greater than for nuclear power. It is not surprising, too, that there are prospects of a world shortage of coal in the long term, say in one or two centuries.

A summary of the prime sources of energy, with their physical requirements of fuel, is drawn up in Fig. 10.1.

10.2.3. Environmental Impact

10.2.3.1 Physical Requirements. Though nuclear and coal-fired systems produce heat in markedly different ways, the 'conventional' plant for using the heat in raising steam to drive turbines to make electricity is similar. It is not surprising that the power stations require roughly the same space. Modern collieries and nuclear reprocessing plants add little to these figures - say 0.1 square kilometres per GW(E)yr in total. However, four times as much area again must be allowed for mine spoil and fly ash dumping over a 30-year plant lifetime. Fusion power stations would, by the same reasoning, require slightly less area per unit output than the nuclear and coal-fired power stations. The turbine exhaust steam from the above processes can be used to supply heat: the combination of heat and power supply (CHP) uses energy about twice as efficiently than if both demands were satisfied by power only. Area requirements are correspondingly reduced.

Tidal power might be said to occupy little area since it merely reduces the shore areas exposed at low tide. On the other hand, the other renewables of wind and wave power occupy considerable areas (see Section 1.4.3) of very roughly 500 and 20 square kilometres per GW(E) installed respectively. The wave machines are, of course, on the sea; in practice, the very large area needed by wind machines implies that much of the siting would also be at sea.

Fig. 10.2 summarizes the above estimates diagrammatically.

10.2.3.2 Waste Arisings and Disposal. There are a number of ways in which nuclear wastes are and will be managed. In order to set out the reference strategy in Section 10.3, it will be assumed that liquid wastes not suitable for pipeline discharge will be converted to solids; further, some LLW will be buried in shallow trenches, whereas ILW and glass HEW will be emplaced in deep tunnels or vaults: the rest of LLW will be placed with the ILW and HLW where convenient. From Table 3.1, ignoring figures for Magnox reactors, which will be phased out by the end of the century, we can derive rough estimates of 2,000 cubic metres for LLW and 300 cubic metres for ILW and HEW (per GW(E)yr). The excavation requirements are thus relatively small (they are, of course, even smaller if spent fuel is buried instead of being reprocessed). Environmentally innocuous methods of disposal can be conceived for most types of nuclear waste (see Appendix 4.4) and the relatively small volumes make associated transport readily feasible. (The disposal methods for Magnox swarf wastes and spent fuel are as yet undesigned.)

In addition to the hazardous materials emitted in low concentrations but appreciable total amounts, and the huge quantities of acid gases and carbon dioxide, coal-fired power produces such large volumes of solid wastes (1.3 million cubic metres per GW(E)yr - see Table 3.2) that they must be dumped (literally) close to their origin, i.e. at mine heads (coal spoil) and at power stations (fly ash) as mentioned in the previous section. Future methods of trapping hazardous materials in stack gases or of improving combustion as in fluidised bed reactors will increase these arisings considerably. Radioactive wastes from fusion are broadly similar in ILW volume to those quoted above for fission power; the conventional wastes from wind and wave power arising mainly during decommissioning (see Table 3.3) amount to about half the LLW figures of Table 3.1. The high figure for tidal power is arguably spurious, since it may be considered unnecessary to remove the barrage (see Section 4.3.2).

Volumes of arisings are shown diagrammatically for each system in Fig. 10.3.

10.2.3.3 Health Detriments - Routine Operation. Table 6.1 summarises the 'routine' death rates for operators in the various power systems: Fig. 10.4 presents the corresponding diagram. Though the estimates are very approximate, the diagram shows how systems needing personnel for many hazardous operations, such as excavating coal or servicing wind and wave machines, have five to ten times the death rate of fission and fusion systems. The latter, because of their relatively high output of power per unit mass of fuel, require fewer personnel, who are, moreover, employed in less hazardous operations.

Table 6.2 gives the corresponding estimates for the deaths among the general public. The system with much the most important effect is that of coal-fired power. Over the short term, this occurs mainly from respiratory diseases. Over a long period (say thousands of years), assuming no improvement in the cure for cancers, the effect of (non-radioactive) cancer-producing chemicals and radioactive emissions contribute roughly the same number of casualties as the (short-term) respiratory effect. This is an important conclusion, since it implies that the long-term hazard of long-lived radioactivity in buried nuclear wastes from any type of cycle is far less than that produced by an equivalent production of coal-fired power. (An improvement in cancer treatment could reduce both long-term hazards.) The stipulation of the Royal Commission on Environmental Pollution 'there shall be no large-scale increase in (UK) nuclear power until long-lived wastes can be satisfactorily disposed' is therefore readily satisfied, when viewed relative to the accepted methods of disposal of wastes from coal-fired power.

10.2.3.4 Health Effects - Large-Scale Accidents. A summary of death rates from large scale accidents for operators and for the general public is given in Table 7.1. The Chernobyl figure dominates the table, being twenty to forty times the level for the Windscale No.1 Pile, Aberfan or coal-mine disasters. The relevant question is 'How meaningful are these past accidents in predicting similar occurrences in the future?' Arguments are presented in Chapter 7 for answering 'not at all'.

10.2.3.5 Overall Death Rates. Using the data established in the preceding sections, overall deaths per unit of power output can be derived. These are set out in the summary diagram of Fig. 10.5. (Not included in these estimates are the potentially immense detrimental effects worldwide caused eventually by the Greenhouse effect and loss of the ozone layer. A drastic reduction of fossil fuel combustion may therefore be advisable on this account alone (McElroy, 1988).) It does not require an expert to see that deaths of the general public due to routine operations of coal-fired power are much greater than the figures relating to Chernobyl. The effects more than a few miles from a UK Chernobyl would be similar to the long-term effects of coal-fired power, except that they would be detectable and would eventually decay away. It is clear, therefore, that whether or not it is justifiable to include Chernobyl (and to a much lesser extent, Windscale No. 1 Pile) in assessing hazards from nuclear power, the steady toxic discharges from coal-fired power are much more hazardous. In both nuclear and coal-fired power, genetic effects are probably negligible. CHP, even if used on (say) 20% of coal-fired power, could only reduce such discharges by about 10%: it does not alter, therefore the general conclusion, on environmental grounds, that coal-fired power should be replaced as rapidly as possible by other power systems. CHP itself then appears less attractive for general application, as the only other power processes with waste heat are nuclear and fusion: these are only really economic in large units. The combination of such large units and the need to site them close to centres of population to allow short transmission distances for heat could then make such CHP systems unpopular.

10.2.4 Financial Aspects

The demand for UK electricity fluctuates daily and seasonally, in addition to a general increase over a period of years. The minimum short-term requirement is known as the base load and the costs for for supplying this by a new PWR, AGR or coal station were compared by CEGB at Sizewell with the average costs of running existing coal stations.

Figure 10.1: ENERGY SOURCES OF POSSIBLE UK POWER SYSTEMS

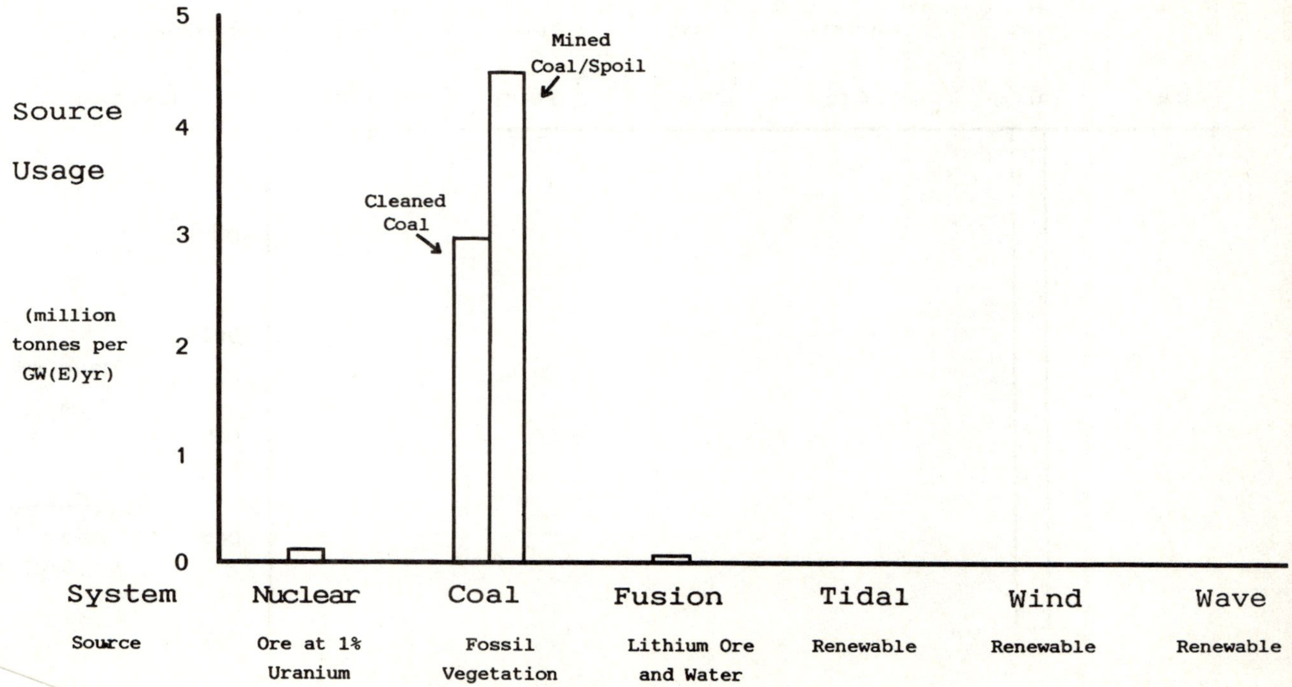

Figure 10.2: AREAS REQUIRED FOR POSSIBLE UK POWER SYSTEMS

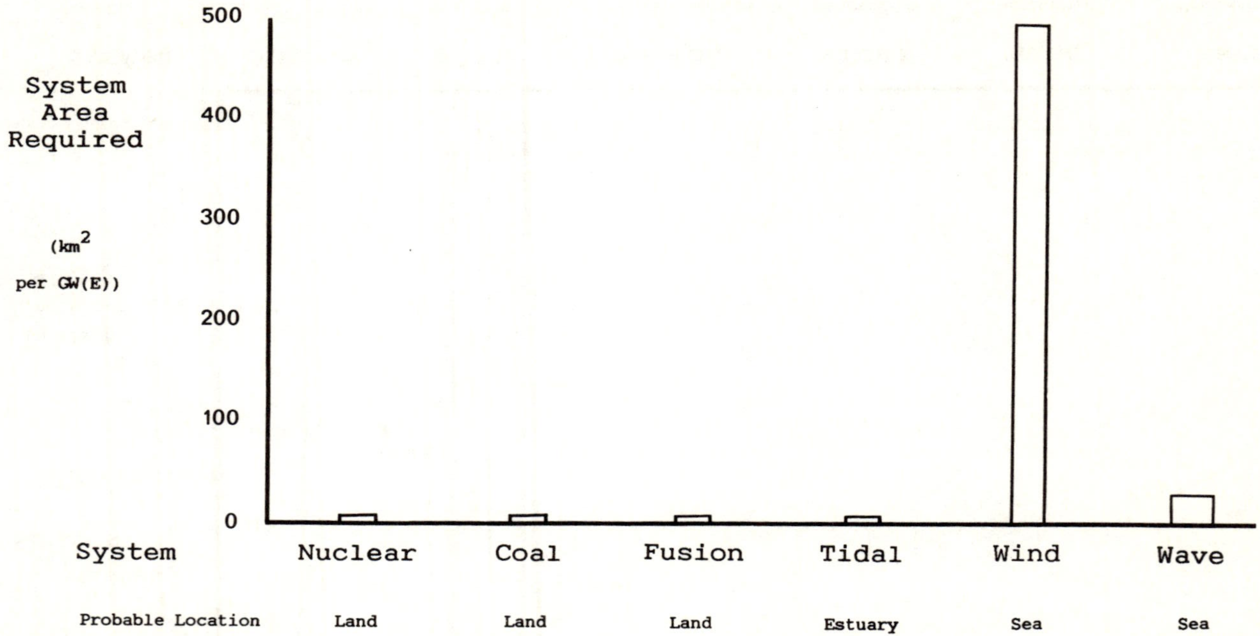

Figure 10.3: VOLUMES OF WASTES ARISING FROM POWER SYSTEMS

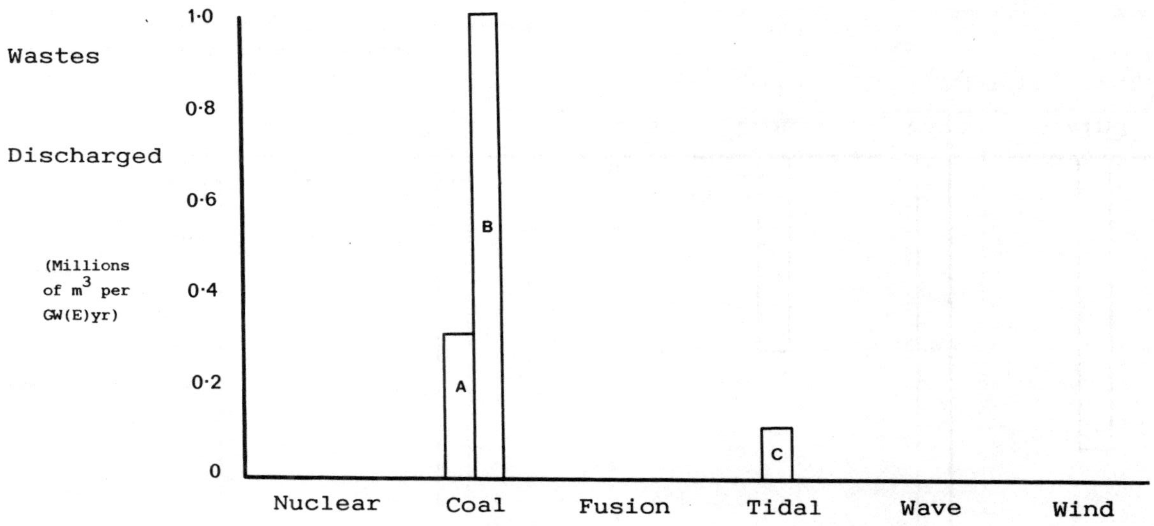

Notes: Figures except A, B and C are too small to be presented.

A represents the volume of fly ash.
B represents the volume of mine spoil.
C represents the volume arising if tidal barrages are dismantled.

Figure 10.4: ESTIMATES OF 'ROUTINE' DEATH RATES OF POWER SYSTEM OPERATORS

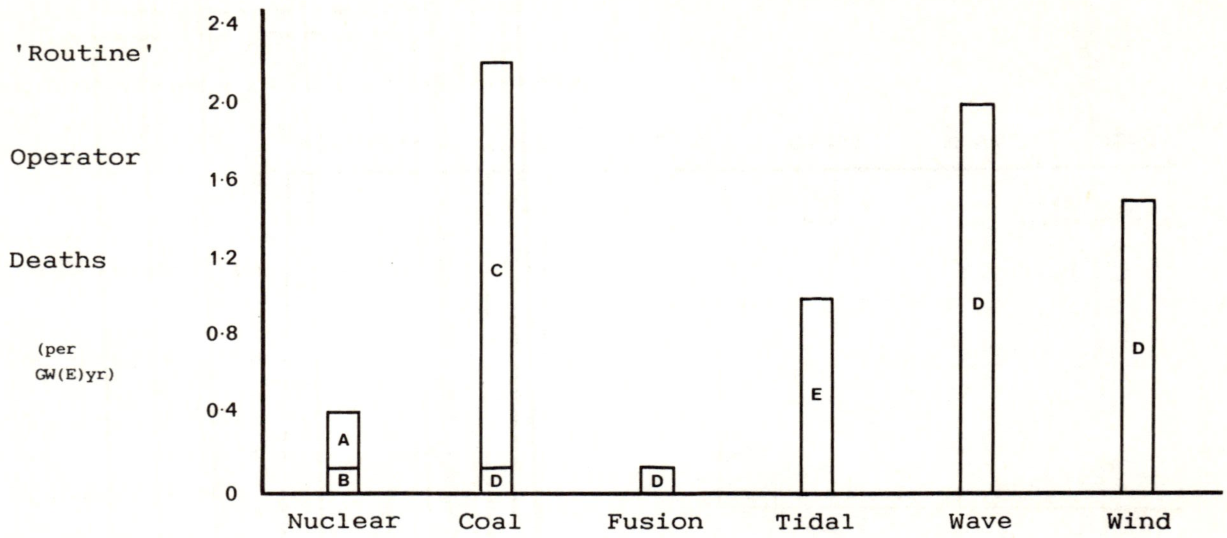

Notes: A represents deaths during reprocessing.
B represents deaths during power station operation.
C represents deaths during mining.
D represents deaths during power station operation.
E represents deaths during construction.

Figure 10.5: OVERA'L DEATH RATES FROM POWER SYSTEMS

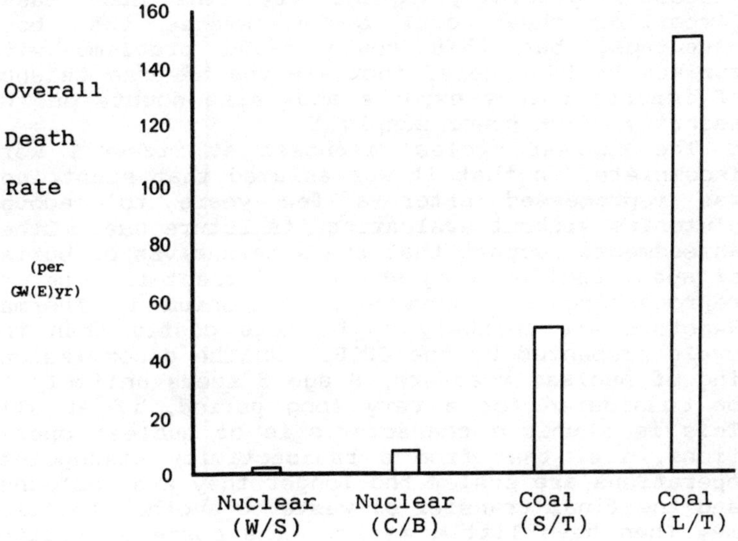

Notes:

Fusion, Wave, Wind and 'Routine' Nuclear and Tidal Power Death Rates are relatively very small. Accidents from Tidal Power are uncertain but could be important.

W/S denotes expected deaths from the Windscale Pile accident averaged over UK Nuclear Power to date.

C/B denotes expected deaths from Chernobyl averaged over world nuclear power to date.

S/T denotes 'short term' deaths from coal-fired power.

L/T denotes 'long term' deaths from coal-fired power.

The clear advantage for nuclear over coal concurs with assessments found in other countries with similar coal extraction costs: the Inquiry Inspector generally agreed with the CEGB case. (Importing cheap coal could reverse the above advantage, but this could cause problems with strikes by UK miners, increase the adverse balance of imports versus exports and raise doubts on the security of UK power supply.)

The nuclear cycles discussed at Sizewell were incomplete, in that it was assumed that spent fuel was reprocessed after a few years to recover plutonium without evaluating its future use. Other assessments suggest that the alternatives of burial of spent fuel or long storage of spent fuel before reprocessing and recycle of plutonium in Thermal Reactors are unlikely to be more costly than the cycle presented by the CEGB. On the decommissioning of nuclear reactors, Stage 3 seems unlikely to be considered for a very long period, if at all. This is a unique characteristic of nuclear operations, i.e. that from a radioactivity standpoint, operations are easier the longer they are postponed and the final transfer of waste to another location may then have little merit. The costs associated with possible Stage 3 decommissioning are therefore also unimportant in comparing nuclear versus coal-fired power. (It is perhaps worth re-emphasising here that though in absolute terms nuclear waste treatment and disposal can cost billions of pounds and justify the search for economies, the cost per kwhr of equivalent electricity is small and unimportant in choosing between power systems.)

It is prudent to place a value on environmental effects, so as to compare them with the plant cost differentials. The analysis of Section 8.4 suggests that effects on artefacts and fauna are monetarily relatively unimportant. Coal power would be at an even higher cost relative to nuclear if human health detriments were included in the comparison, especially if the valuation of life were consistent with that used in extra cleanup of Sellafield discharges. Overall, preference of a PWR over a new coal station the size of Sizewell could save each year about 50 deaths and pay for 6,000 hip replacements or 1,500 kidney transplants or several new hospitals. A further (as yet uncosted) disadvantage in the long-term future is the Greenhouse effect from the combustion of coal which could have enormous worldwide consequences.

Currently, Fast Reactors appear less economic than PWRs (Section 8.3.3), but in the long-term future, beyond that reviewed at Sizewell, fossil fuels will begin to get more difficult to extract; pressure on uranium supplies will increase and associated costs will rise. By then, Fast Reactors appear to be the most promising source of power, at least in countries with advanced lifestyles demanding power of high availability and reliability.

10.2.5 Political Features

With the decreasing urgency to develop new reactor systems, in particular commercial Fast Reactors, the UKAEA is shrinking and becoming less important in nuclear matters. The two British Companies, NNC and BNFL, formed to compete with large overseas companies in reactor-building and fuel cycle operations, effectively have monopolies in the UK, and so nuclear costs have tended to rise closer to those of coal. NNC and BNFL thus increase their income and profits at the expense of higher electricity prices, i.e. they charge through CEGB what the market will bear.

The Tory Government remains in favour of increased nuclear power, though an important part of the driving force has diminished with the decreasing influence of the NUM after its defeat in the 1984 miners' strike. On the other hand, the Government plans for power privatisation reduce the cost advantage of nuclear over coal because the private investor will require a higher interest rate than the State. On nuclear waste, successive Governments have carefully adopted a laid-back stance in order to avoid involvement in decisions which can only be unpopular. Opposition parties have adopted unrealistic and inconsistent attitudes, mainly against nuclear power; these are clearly intended to be vote-catching but impractical to carry out if such parties ever formed a Government.

In spite of the above differences in the attitudes of political parties to nuclear power and clear requirements in their constitutions to avoid political bias, the existing teloids, the BBC and IBA, have behaved as members of the holoid anti-nuclear chorus. Careful censorship of pro-nuclear facts is coupled with instantaneous broadcasting of anti-nuclear information. The latter appears at the head of items at main news times, to be faithfully followed by front-page reiteration in the next morning's newspapers (see Figs 9.2 and 9.3). Though measures to encourage the media to be less parsimonious with the truth may well be developed, political deals may dilute their efficacy. It is prudent therefore to assume, in designing systems for the future, that alarmist anti-nuclear campaigns will persist.

Though anti-nuclear campaigns by the media and the 'Greens' occur in many countries, the desire of the populations of more advanced countries to maintain their sophisticated lifestyles with cheap electricity always 'on tap', together with the increasing worldwide demand to reduce fossil fuel combustion and its stack discharges, will result in an increasing need for nuclear power. The resulting competition for uranium will increase its price so that eventually Fast Reactors will become economic and form part of the nuclear power supply. In the short term, however, say for twenty or thirty years, the UK power policy will probably follow the political flavour of successive Governments and be only slightly affected by developments in other countries.

The decision to replace coal power by operating a Once-Through nuclear cycle is straightforward on environmental grounds, even though only a crude historical approach has been used in our comparisons. Having reached this conclusion, the secondary question to be addressed is where are the best locations for the power stations and waste disposal. (The volume and costs of wastes from future Once-Through Cycles should be carefully distinguished from the much greater accumulations from reprocessing cycles, either to date or in the future.) Though measures may be taken to curb distortion of the truth by the media, its effect to date on the public has been to instil a long-term worry over the safety of nuclear operations. The impracticability of evacuation plans after an incident on nuclear plant has been highlighted by Openshaw (1986), who has therefore suggested siting such plant more remotely from large centres of population. It would be prudent, therefore, to consider the implications of siting if a large UK nuclear power programme eventually evolved. The need for the decommissioning of nuclear plant to include Stage 3 is also clearly reduced by a 'remote siting' policy. Though the CEGB has steadily maintained that the likelihood of severe reactor accidents in the UK was acceptably low, so that remote siting was unnecessary, it will be interesting to see whether they allow their original technical judgements to be overridden by public opinion, rather than openly admit to the practical sense of remote siting. (Bowing to public pressure is, of course, a technique already being used by NIREX instead of insisting on its own technically preferred options.)

Clearly, from our earlier comparisons of the hazards from nuclear and coal-fired systems, safe disposal of ILW and HLW is technically feasible in several broadly different ways. However, disposal at sea is operationally less easy than from land; further, the political 'mileage' to be gained by other countries objecting to sea disposal probably outweighs the needs of excavation for burial underground. The simplest form of the latter (USL as defined in Section 4.1.8.1)) is to bury the waste close to where it is produced, thus reducing packaging costs to meet transport requirements on rail and roads and arousing public opposition en route. This 'KISS' (Keep-It-Simple-Stupid) approach in practical terms currently implies burial at Dounreay and Sellafield, as proposed about ten years ago by the author. The methods most simple and least hazardous to construct are to locate the wastes above the water table in Dry Boxes, or Dry Repositories, with drainage to sea. Alternatively, sufficient safety will be achieved by locating disposal in stagnant saline groundwater under the sea-bed. The local geology will be of secondary importance when the impact of extremely slow nuclide migration to sea is compared with the leaching of fly ash and mine spoil wastes and the general toxic cocktail emitted from coal-fired power station stacks. Moreover, as mentioned in Appendix 6.2, hazards of transport over long distances can be greater than benefits postulated from 'better' geology. The value of continuing sophisticated research, especially into inland sites using complex multi-barrier designs, is then questionable. The acceptance of this simple approach may come, not merely because of its basic commonsense, but from the rundown of nuclear fuel cycle operations at Dounreay and Sellafield, raising the spectre of high unemployment for nuclear operatives.

The question of early recovery of plutonium from spent fuel is of much lower importance than the basic choice of nuclear power as the main replacement of coal. The need for plutonium in UK Fast Reactors is clearly not urgent: on the other hand, reprocessing has only slight environmental impact (including possible accidents) and complete shutdown of operations is commercially difficult because of the Japanese contract. A reasonable assumption is that all the remaining UK Magnox fuel should continue to be reprocessed early in B205 at Sellafield and THORP run at an economic output level. Possible excess requirements over this could be accommodated by building long-term stores for AGR and PWR fuels. Plutonium stocks could be controlled by using MOX fuels, at least in PWRs.

Concerning fusion and renewable systems, it seems reasonable to suppose that if and when fusion reactors are economically viable, they will begin to replace fission reactors; corresponding siting requirements will be very similar and so fitting in fusion stations should be straightforward. A minor fraction of UK power (say 20%) could be foreseen to come from renewable systems; their location would be determined largely on their technical characteristics and availability of suitable land (or sea). UK Government policy is to assist in the development of renewables until preferred systems are identified in the 1990s. Conservation and CHP applications may help to give useful minor reductions of power requirements, but would not affect the case for providing the bulk of electricity from nuclear fission and not from combustion of coal.

10.3 A REFERENCE SYSTEM FOR UK POWER

With the background of technology broadly presented in this book and the need for flexibility to cater for the vagaries of the future, we will set out below one possible scheme for future UK power.

Fig. 10.6 shows an outline of the UK, in which shaded areas represent broadly the regions of major electricity consumption. Some representative centres of future power are also shown and described below with more detailed illustrations in Figs 10.7 and 10.8.

G1 indicates a possible area of development for geothermal energy, assuming a successful outcome to the research described in Section 1.4.3. Tidal schemes T1 and T2 are indicated in the Severn and Mersey Estuaries.

C1, C2 and C3 are typical coastal sites of nuclear power generation: currently, Sizewell (C1) has been sanctioned as a site for a PWR, and CEGB have indicated that Hinkley Point (C2) in Somerset and Wylfa (C3) in Anglesey are also probable PWR sites. Other existing nuclear sites, which may be proposed for PWRs as Magnox reactors are phased out, are omitted for clarity. It is suggested here that 'C' class sites would be be confined to existing sites in areas of low population density and would 'burn' mainly uranium fuels. These fuels, when spent, would be transported off-site for eventual reprocessing or long-term storage. FS1 and FS2 denote long-term spent fuel storage and/or waste disposal facilities at Dounreay and Sellafield.

As the nuclear programme expands, remote sites will be desirable. PC1 denotes a 'Power Complex' on a man-made island of the 'fill' type on Long Sand in the Wash (mentioned in Sections 4.1.7 and 4.1.8.4.). Such complexes are convenient for nuclear plants which, if required, will be charged with plutonium-enriched fuels: reprocessing and fabrication plants can be installed as necessary to receive spent fuel from 'C' class sites and also spent fuel from their own or other 'PC' sites. Since 'PC' sites will have their own harbour, it will be possible, if so desired, to transfer fuels between them entirely by sea. Some transfers of plutonium recovered by reprocessing at Dounreay or Sellafield might be useful from their nearby harbours to balance plutonium requirements in the medium term.

The siting of 'PC' complexes has been arrived at by considering the attractiveness of remote sites and avoiding excessive transmission losses over long distances to the main centres of electricity consumption. Advantage has been taken of the natural features of sandbanks for constructing the preferred form of island - the 'fill' island: the depth and structure of the fill may be dictated by foundation requirements for the heavy loading per unit area caused by reactor structures. Apart from the obvious reduction in fill in raising sandbanks compared with building from the sea-bed in relatively deep water, tidal flows round the sandbank will not be altered markedly, so that possible deleterious effects from erosion on the mainland due to alteration of currents can be minimised. Other 'PC' sites are indicated in Morecambe Bay (PC2) and on Buxey Sand in the Thames

Estuary (PC3). If developments in cheapening transmission, such as the use of ceramic superconductors, prove successful, more remote sandbanks might be considered for PCs, e.g. off the Firth of Forth (PC4) and the Norfolk coast (PC5). Transmission from mainland areas with sparse population, such as the North of Scotland, are not suggested, as they would be prone to the vagaries of nationalist politics.

The postulated complex on Long Sand is shown in more detail in Figs 10.7, 10.8 and 10.9. A natural attraction of this sandbank is that relatively deep water exists on either side in Boston and Lynn Deeps. One obvious physical layout of the complex is to follow the length of Long Sand, with cooling water intakes on one side, say the Boston Deep, so that the water, warmed after use in reactor condensers, would be discharged on the other side, say in the Lynn Deeps, avoiding mixing with the intake water, which would reduce the efficiency of power production (see plan view in Fig. 10.7). (Apart from the daily tidal flows in and out of the Wash, there is a net flow in the adjacent deep sea which eventually takes warmed water away.)

The abrupt increase in depth at the edge of the sandbank near Lynn Deeps suggests a possible location for a harbour. From the mainland, either a bridge (low cost) or solid embankment (fitting into later barrage schemes) could be chosen for access. (The embankment would probably not be selected if such barrages (see later) were shown to be undesirable, since erosion of coasts from changes in tidal currents could result.) Some uranium-burning reactors (UR) might be initially installed, with the harbour and appropriate service areas for construction, equipment and spent fuel storage. Later, plutonium recycle reactors (PR) (either Thermal or Fast) would follow, eventually with their own fuel reprocessing and fabrication plants. Alternatively, if fusion power (FP) became attractive, units could easily be fitted into the general layout of the cooling and transmission systems. An interesting characteristic of the island construction is that the sea defences are a major component of its cost, so that a small increase in width can give an increase in enclosed area for a less than proportional change in island cost. Areas can then be left, designated for other reactors or (say) waste disposal; very low-activity wastes, say from the demolition of mainland reactors, could be used as general fill, foundations of new plant or in new sea defences. Excavation to greater depths within the island, preparatory to LLW or ILW disposal, could provide 'fill' for raising levels elsewhere on the island. Alternatively, both LLW and ILW could be disposed within shells of decommissioned reactors and covered, if desired, with clay and soil to form a version of a Dry Box (Section 4.1.7). Shafts could be sunk deep into the clay and evaporites beneath the site for disposal of HLW and spent fuel in saline groundwater zones in the manner described in Section 4.1.8.4. (Much preparatory geological information for the area was obtained during the feasibility study of the proposed Wash Barrage a couple of decades ago.)

A futuristic and perhaps fanciful eventual development of the site to perhaps 10 to 20 GW(E) capacity is illustrated in Fig. 10.9. Here, the geometry of the site and its land connections can divide water flows so as to allow a series of tidal power schemes to be installed. It is well known that concepts of a Wash Barrage for power and/or freshwater storage have been difficult to set out in a single stage, because of the enormous volumes of water moving with each tide: the closing of final sections of a barrage were then subject to prolonged and concentrated flows. In Fig. 10.9, solid access routes to the island could be used in the stagewise reduction of water flows. Some enclosed segments could then be designated for freshwater storage and others for tidal power. The sheltered waters behind the barrages could facilitate the erection of wind generation units, while wave power units could be stationed seaward of the outer barrage.

Overall, therefore, it is possible to set out a future power scheme for the UK with many desirable features - flexibility, remote siting, disposal of wastes near main sources, low environmental impact, etc. However, the reader by now will have sufficient know-how to put together other combinations of power systems. Maybe we shall yet see some of those on 'Tomorrow's World'! But don't be too hopeful; even if you manage to pass the ball to a programme director and he doesn't drop it, he may not run in the right direction!

Figure 10.6: POSSIBLE IMPORTANT SITES FOR FUTURE POWER IN GREAT BRITAIN

(For Interpretation of Symbols, see Section 10.3)

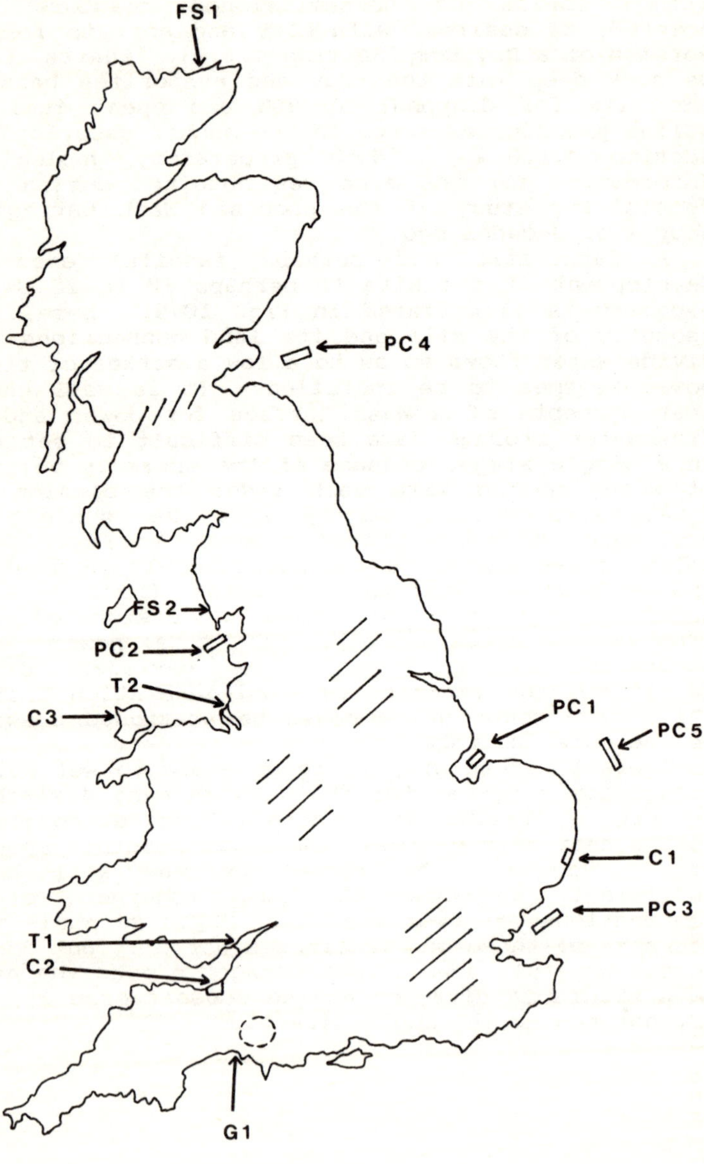

Figure 10.7: WASH POWER COMPLEX

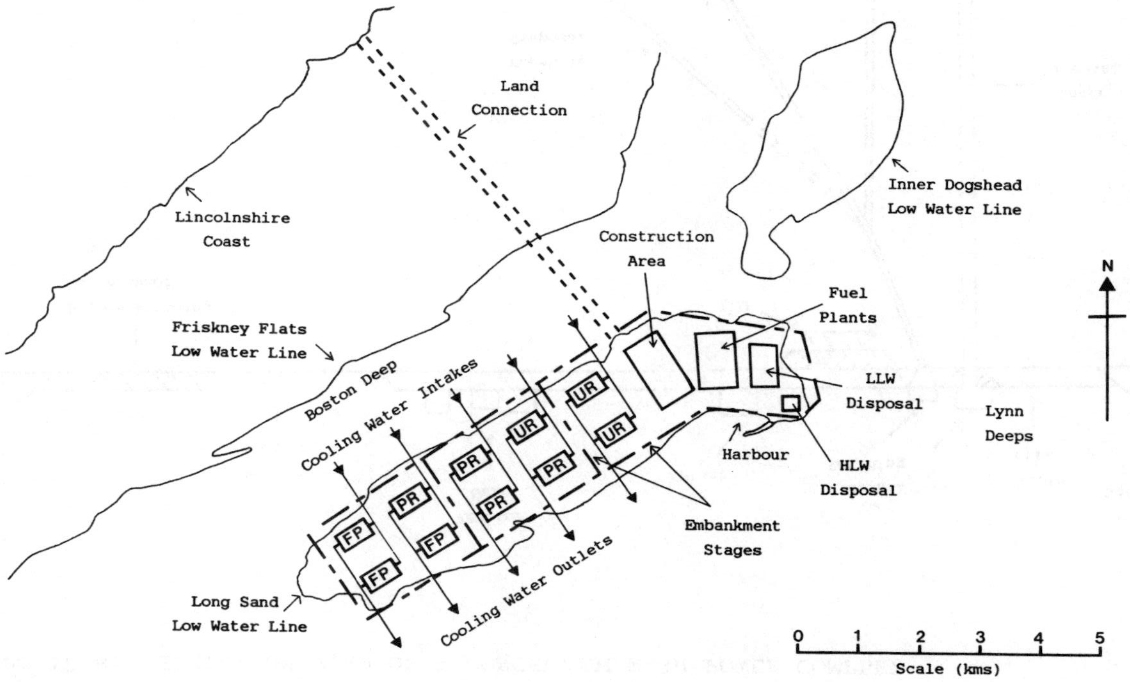

Figure 10.8: ELEVATION VIEW OF A CONCEPTUAL WASH POWER COMPLEX

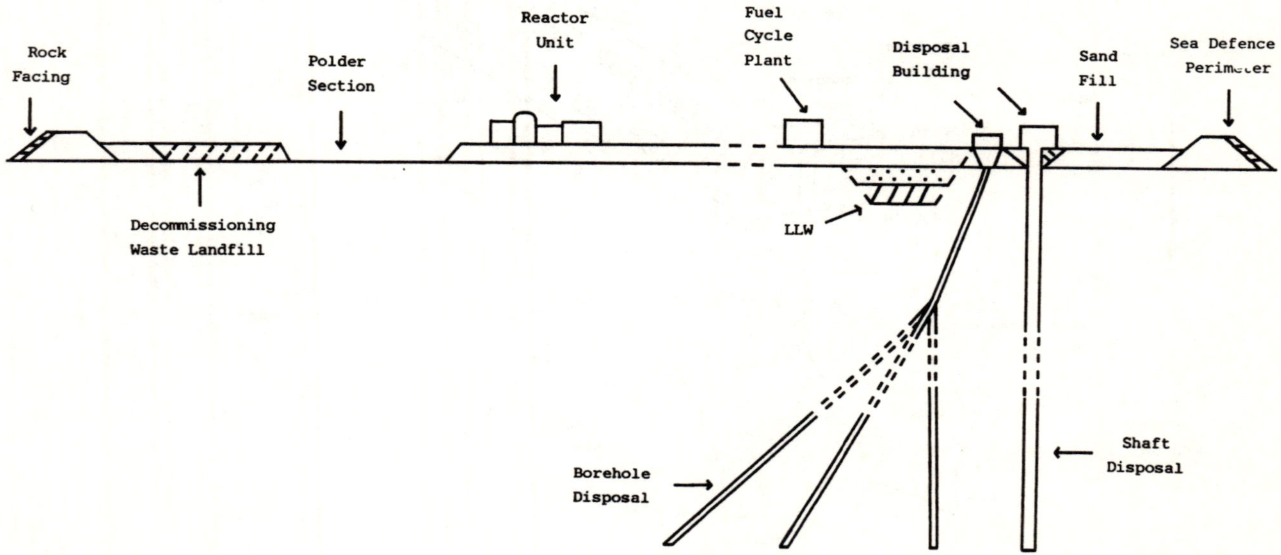

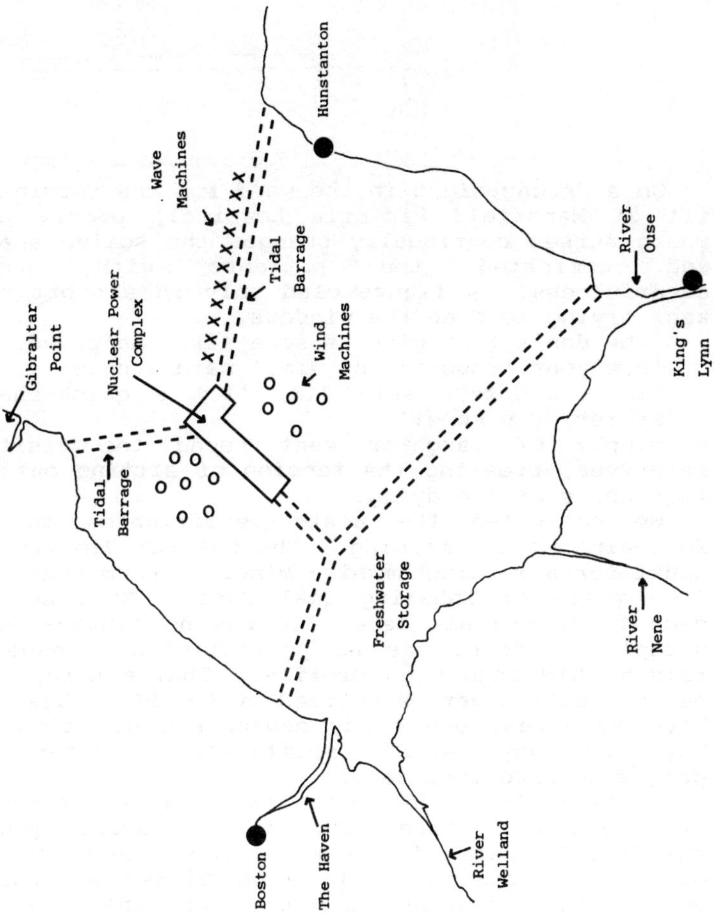

Figure 10.9: WASH POWER COMPLEX - LONG-TERM POWER DEVELOPMENT

EPILOGUE

On a Friday night in the ward for the terminally ill at Mansfield Victoria hospital, poorly paid young nurses continually changed the soiled sheets and comforted the patients with heroic cheerfulness. A figure clad only in a short vest kept trying to open the windows.

'Why don't they give us some air,' he cried.

'Put your trousers on, Jim,' said a nurse.

'In a minute,' said Jim, 'I'm a quick-change artist, you know!'

A ripple of laughter went round the visiting relatives, breaking the tension of sitting out the last hours of the dying.

We collected the death certificates on the following Monday morning. The one for Jim was the usual for a Nottinghamshire miner - pneumoconiosis, from years of inhaling coal dust. No reporters came to record his case; Jim was no 'three-legged sheep', so there was no journalistic mileage in raising him into high profile. That's Life, Jim, but you will never be noticed by the BBC. Dead men have no votes, don't buy newspapers or watch the box, so why should politicians, editors or programme directors care?

So life, in the UK, with its own live version of 'Yes, Editor', rolls on. Without nuclear power, the fossil fuels of the world would begin to run out in the next century, with bitter scrabbling among the advanced nations over the remains. Thermal Reactors provide a breathing space and Fast Reactors a much longer one, fuelled by one of the Earth's oldest inhabitants from supernova days, plutonium. Will Brand X, a wonderful new source of power, cheap, in endless supply and innocuous, ride in on a white charger to save mankind from the Greenhouse effect and the supposed catastrophies of the proliferation of nuclear power? Will the Friends of Fission (FOF) then rise up and oppose it, aided and abetted by the then champion of lost causes, Channel 13? Will the Friends of Fusion (FOF) form an indistinguishable splinter group?

However, as the chat shows increase to fill the time available on ever more television channels, and new pressure groups debate at length the trivial issues they feel they understand, one may be sure that Parkinson's Laws will survive, albeit in new guises.

REFERENCES

Allott and Lomax, Consulting Engineers, Sale, Manchester, Private Communication, November, 1986.

Armstead H.C.H., *Geothermal Energy*, pp.286-94 (E. and F.N. Spon, 1983).

Ashley H., Rudman R.L. and Whipple, C. (eds), *Energy and the Environment: A Risk Benefit Approach* (Pergamon, 1976).

Barret L. and Waddell T., *Cost of Air Pollution Damage* (US Environmental Protection Agency, 1973).

Barrett G. W., 'Methods for reducing emissions to air from power stations', *Clean Air*, Vol.16, No.1, p.3 (1986).

Batchelor A.S., 'Progress in hot dry rock exploitation', *Energy Research*, Vol.9, pp.377-90 (1985).

BBC2, The Dump, 25 November (1986).

BBC2, Taming the Dragon, Brass Tacks Special, October 1987.

BBC2, The Greenhouse Effect, 8 February (1988).

Beckmann P., *The Health Hazards of Not Going Nuclear* (Golen Press, 1976).

Bengtsson G., Comparison of Radiation and Chemical Risks, SSI-rapport 88-18 (1988).

Binnie and Partners, *Islands for Offshore Nuclear Power Stations*, EUR 7534 (Graham and Trotman, 1982).

Birt J., Fleming Memorial lecture to the Royal Television Society, April 1988.

Black Sir D. *et al.*, *Investigations of the Possible Increased Incidence of Cancer in West Cumbria etc.* (HMSO, 1984).

Black S.C. and Niehaus F., 'How safe is 'too' safe', *IAEA Bulletin*, Vol.22, No.1, p.40 (1980).

Blowers A. and Pepper D. (eds), *Nuclear Power in Crisis* (Nichols, Croom Helm, 1987).

BNFL, Annual Report on Radioactive Discharges and Monitoring of the Environment (1986).

British Coal, *Clean Use of Coal* (British Coal Corporation, 1987).

Broecker W.S., 'Unpleasant surprises in the greenhouse?' *Nature*, Vol.328, pp.123-26 (1987).

Burton W.R., 'Shielded containers for waste could prove beneficial', *Nucl. Eng. Int.*, Vol.28, No.323,p.41 (1982).

Burton W.R., Hazardous Waste Disposal, GB Pat. 2166282.
Burton W.R. and Cairns W.J., British Radwaste: is a long journey really necessary?, *Nucl. En. Int.*, p.22 (May 1983).
Burton W.R. and Griffin J.R., A Design Study of Long-Term Storage and Underground Systems for Highly Active Waste, ND-R-514(R), (1981).
Bury M.R.C., The Offshore Disposal of Radioactive Waste by Drilled Emplacement, EEC Contract 256-81-7 WAS UK (1985).
Camplin W.C., BNFL Sellafield, Assessment of Public Radiation Exposure etc., Sizewell Inquiry Series No.9 (1982).
Camplin W.C. and Hill M.D., 'Sea dumping of solid radioactive waste, etc.', *Rad. Waste. Mgt. and Nucl. Fuel Cyc.*, p.233, Vol.7, No.3 (1986).
CEC, Methodology for Evaluating the Radiological Consequences, etc., Commission of European Communities., Doc.No.V/3865/79-EN FR (1979).
CEGB, *Analysis of Generating Costs*, 1983/4 Update (CEGB, 1985).
Chappie M. and Lave L., 'The health hazards of air pollution: a reanalysis', *J. Urb. Econ.*, 12, 346 (1982).
Chudleigh R. and Cannell W., *Radioactive Waste: The Gravedigger's Dilemma* (FOE, 1984).
Clarke R.H., Environmental Impact of Nuclear Generating Stations, London Conference, 26 April, 1982.
Cohen A.V. and Pritchard D.K., 'Comparative Risks of Electricity Production Systems, A Critical Survey of the Literature', HSE Research Paper No.11 (1980).
Cohen B.L., 'Long-term consequences etc. for chemical carcinogens', *Risk Analysis*, Vol.1, No.4, p.267 (1981).
Cohen B.L., *Before It's Too Late*, (Plenum Press, 1983).
Collier J.A. and Davies L.M., *The Accident at Chernobyl Unit 4 etc.* (CEGB, 1986).
Comar C.L. and Sagan L.A., *An. Rev. of Energy*, Vol.1 (1976).
Combes B. and Slovic P., 'Newspaper coverage of causes of death', *Journalism Quarterly*, 56(4), pp.837-43 (1979).
Commission on Energy and the Environment, *Coal and the Environment* (HMSO, 1981).
Cottrell A., *How Safe is Nuclear Energy*? (Heinemann, 1981).

Craft A. and Openshaw S., 'Children, radiation, cancer etc.', in Blowers and Pepper (eds), *Nuclear Power in Crisis*, p.244.
Crick N.J. and Linsley G.S., *Assessment of the Radiobiological Impact of the Windscale Reactor Fire*, NRPB-R135 and Addendum (HMSO, 1983).
Curran J. and Seaton J., *Power Without Responsibility*, (third edition, Fontana, 1988).
Currie W.M., 'Chasing the energy efficiency goal', *ATOM*, No.382, August 1988.
Cutler J., Dispatches Programme, ITA Channel, 4 March 1988.
Dart J.F. and Talbot J.R.W., 'The CEGB's role in CHP development', ATOM, No.342, p.9 (1985).
Davis J.P. and Smith G.M., *Radiological Aspects of Solid Wastes from D-T Fusion Reactors*, NRPB-R210, (HMSO, 1987).
Day J. and Hodgson D., 'Pollution control: the cases of acid rain and lead in petrol', *Economics*, Autumn, p.106 (1985).
DOE, *Co-operative Programme of Research on the Behaviour of Hazardous Wastes in Landfill Sites* (HMSO, 1978).
DOE, *Assessment of Best Practical Environmental Options (BPEOs) for Management of LLW and ILW Solid Radioactive Wastes* (HMSO, 1986).
DoEn, *CHP and Electricity Generation in the UK*, Paper No.35 (HMSO, 1979).
DoEn, Proof of Evidence to the Sizewell B Enquiry, October, 1982.
DoEn, *Privatising Electricity* (HMSO, 1988a).
DoEn, *Renewable Energy in the UK: The Way Forward*, Paper No.55 (HMSO, 1988b).
Elayi A.G. and Schapira J.P., 'Long-term radiotoxicity of high level wastes etc.', *Rad. Waste Mngt. Nucl. Fuel Cycle*, Vol.8(1), p.1 (1987).
Elkington J., *Sun Traps* (Penguin Books, 1984).
Environmental Sciences Inc., Land Improvement with Waste Materials, UK Pat. Spec. 1,337,301 (1973).
Ewart F.T., The Comparative Hazards of Chemicals and Radioactive Waste, AERE R.11105 (1983).
Fairhall D., 'Why you can't get away from it all', *Guardian*, 6 June (1986).
Farmer A.A., 'Recycling of fuel', *Nuclear Power Technology*, Vol.2 (Oxford, 1983).
Fishlock D., 'Public perceptions on nuclear waste', *Rad. Waste Mgt.*, p.25 (BNES, 1984).

Fletcher A.M. *et al.*, The 1985 UK Radioactive Waste Inventory, DOE/RW/86/087 (1986).

Flood M., *Energy Without End* (FOE, 1987).

Flowers R.H., British Nuclear Energy Society Conference on Radioactive Waste Management (BNES, 1984).

Foley G. and von Buren A. (eds), *Nuclear or Not, Choices for Our Energy Future* (Heinemann, 1978).

Forbes I.A. *et al.*, The Nuclear Debate, A Call to Reason, Energy Research Group, Boston (1974).

Forman D. et al., 'Cancer near nuclear installations', *Nature*, 8 October (1987).

Freeman T.J. *et al.*, 'Modelling radioactive waste disposal etc.', *Nature*, Vol.310, No.5973, pp.130-3 (1984).

Fremlin J.H., *Power Production, What Are the Risks?* (Hilger, 1985).

Fremlin J.H., 'The hazards of civil nuclear power', *Sci. Parl.*, Vol.43, No.185, p.221 (1986).

Gardner M.J. *et al.*, Follow Up Study etc., Brit. Med. J., Vol.295, p.819 (1987).

Gittus J.H. *et al.*, *The Chernobyl Accident and its Consequences*, NOR 4200 (HMSO, 1987).

Greenhalgh G., *The Necessity For Nuclear Power* (Graham and Trotman, 1980).

Greenpeace, 'Britain's shame', Greenpeace News, October 1986.

Gregory J.W., 'A deep bore at Seascale in Cumberland', *Geol. Mem.*, p.146, December (1915).

Griffin J.R. *et al.*, An Engineering Design Study for Storage and Disposal of Intermediate Level Waste, ND-R-777(R), (1982).

Grove-White R., *Energy Self-Sufficiency for the UK?* p.151 (Gower, 1985).

Hamilton L.D., Testimony At Sizewell B Public Inquiry, APG/P/4 (1983).

Hamilton L.D. and Manne A.S., *Nuclear Power and its Fuel Cycle*, Volume 7, pp.196-214, IAEA, Vienna (1977).

Hancox R., Towards Fusion Generation: Inst. Nucl. Eng. Lecture, London (1987).

Haslam C.J., 'The Rationality of Corporate Planning Techniques', Ph.D. Thesis (Aberystwyth, 1988).

Hawkes N. *et al.*, *The Worst Accident in the World* (Pan Books, 1986).

Health and Safety Commission, *The Hazards of Conventional Sources of Energy* (HMSO, 1978).

Health and Safety Executive, The Leakage of Radioactive Liquor into the Ground, BNFL, Windscale, 15 March 1979 (1980).
Health and Safety Executive, Windscale: The Management of Safety (1981).
Henderson M., *Living With Risk, The British Medical Association Guide* (Wiley, 1987).
Hill M.D. and Grimwood P.D., *Preliminary Assessment of Radiological Aspects of Disposal etc.*, NRPB-R69 (HMSO, 1978).
Hill M.D. and Lawson G., *An Assessment etc., of High Level Waste in Coastal Geological Formations*, NRPB-R108 (HMSO, 1980).
Hill M.D. *et al.*, *An Assessment etc. of Disposal of Intermediate Level Wastes in Argillaceous Rock Formations*, NRPB-R126 (HMSO, 1981).
Hill Sir J., 'New technologies in a changing energy situation', *ATOM*, No.319 p.1 (1983).
Howarth G.G., 'Encapsulation of ILW by BNFL', paper presented to Waste Management '87, Tucson, Arizona (1987).
Howden M., *Progress in the Reduction of Radioactive Discharges from Sellafield* (BNFL, 1988).
Hull A.P., 'Radiation in Perspective etc.', *Nuclear Safety*, Vol.12, No.3, May-June 1971.
ICRP, *Limits for Intakes of Radionuclides by Workers*, Publication 30 (Pergamon, 1981).
ICRP, *The Metabolism of Plutonium and Related Elements*, Publication 48 (Pergamon, 1986).
Ince M., *Sizewell Report, What Happened at the Enquiry?* (Pluto Press, 1984).
Independent Television Programme, The Acid Test, 25 October (1986).
Inhaber H., *Energy Risk Assessment* (Gordon and Breach, 1982).
Jones P.M.S. (ed.), *Nuclear Power: Policy and Prospects* (Wiley, 1987).
Katz J.J., Seaborg G.T. and Moss L.R. (eds), *The Chemistry of the Actinide Elements*, second edition, p.503 (Chapman and Hall, 1986).
KBS, Final Storage of Spent Fuel, KBS-3 (1983).
Kinlen L., reported in the Times, 9 December 1988.
Knowles A.N., *Radioactive Waste Management, Technical Hazards and Public Acceptance*, p.85 (Oyez Scientific Technical Services, 1985).
Lee T.R., *Radioactive Waste Management, Technical Hazards and Public Acceptance*, p.213 (Oyez Sci. Tech. Serv. 1985).
Longhurst J.W.S., *Environmental Implications of the UK FGD Programme* (Manchester Polytechnic, 1987).

McElroy M., 'The challenge of global change', *New Scientist*, p.34, 28 July 1988.
McKerron G., 'Is Sizewell a good investment?', *Public Money*, p.3, December 1984.
Milborrow D.J., 'Wind energy research in the CEGB', *ATOM*, No.344, p.6 (1985).
Murray C.N., Barbreau A. and Burdett J.R. (in) Radioactive Waste Management and Disposal, EUR 10163 (1986).
Murray C.N., Weydert M. and Freeman T., 'An ocean-satellite link etc.', *Deep Sea Research* (in press) (1987).
National Coal Board, Annual Report for 1984-5 (1985).
NCB, *Fluidised Bed Combustion*, second edition (National Coal Board, 1985).
Novegno A. and Asculai E., 'In perspective; the role of safety assessment etc.', *IAEA Bulletin* 2, p.33 (1987).
Nuclear Technology (Consultants), Long Thimble Computer Programme (1986a) (Unpublished).
Nuclear Technology (Consultants), Diffusion Computer Programme (1986b) (Unpublished).
O'Brien B., '*What Are My Chances, Doctor?*' Office of Health Economics, Paper 84 (OHE, 1986).
OHE, Measurement of Health, Paper 77 (Pub. by Office of Health Economics) (1985).
Openshaw S., *Nuclear Power, Siting and Safety* (Routledge and Keegan Paul, 1986).
Openshaw S. *et al*, 'Investigation of leukaemia clusters etc.', *The Lancet*, Vol.I (1988).
Ove Arup and Partners, *Ocean Disposal of Radioactive Waste etc.*, p.104 (Graham and Trotman, 1985).
Ozkaynak H. and Spengler J.D., 'Analysis of health effects resulting from population exposures to acid precipitation precursors', *Env. Health Persp.*, Vol.63, pp.45-55 (1985).
Parker R., *The Windscale Inquiry* (HMSO, 1978).
Parkinson C.N., *Parkinson's Law or The Pursuit of Progress* (Murray, 1958).
Passant F., *Radioactivement Waste Management, Technical Hazards and Public Acceptance*, p.65 (Oyez Scientific Technical Services, 1985).
Patterson W.C., *Nuclear Power* (Pelican Books, 1983).
Patterson W.C., *Going Critical* (Paladin, 1985).
Patterson W.C., Lies, damned lies and Magnox, *Guardian*, p.28, 11 October, 1988.

Pearce D., The role of economics in assessing desirable air quality levels, *Clean Air*, Vol.13, No.3, p.88 (1983).
Pearce F., *Acid Rain* (Penguin Special, 1987).
Pickaver A., *The Pollution of the North Sea* (Greenpeace, 1981).
Pochin E., Estimated Population Exposure, Nuclear Energy Agency, Paris (1976).
Porritt J., *Seeing Green* (Blackwell, 1984).
Price B., *Guide To Pollution* (Temple Smith, FOE, 1983).
Pryke R., 'Privatising electricity supply', *Fiscal Studies*, Vol.8, No.3, pp.75-88 (1987).
Radioactive Waste Management Advisory Committee, Sixth Annual Report (HMSO, 1985).
Richards H.J., *Radioactive Waste Management, Technical Hazards and Public Acceptance*, p.117 (Oyez Scientific Technical Services, 1985).
Ringwood A.E., *Safe Disposal of High Level Nuclear Reactor Wastes etc.*, (Canberra, A.N.U. Press, 1978).
Roy R., *Radioactive Waste Disposal*, Vol.1 (Pergamon, 1982).
Royal College of Nursing, *Major Joint Replacement: Nursing Guidelines* (Harrow Scutari Projects, 1988).
Royal Commission on Environmental Pollution, 'Nuclear Power and the Environment', RCEP Sixth Report (HMSO, 1976).
Royal Commission on Environmental Pollution, 'Tackling Pollution - Experience and Prospects', RCEP Tenth Report (HMSO, 1984).
Seymour A.H. (ed.), *Radioactivity in the Marine Environment* (Nat. Acad. Sci., 1971).
Shimmin F.N., 'The engineering of effluent disposal to sea at Dounreay and Winfrith', *J. Inst. Pub. Health Eng.*, 62:181-204 (1963).
Taylor R. and Wilkie D., 'Drawing the line with leukaemia', *New Scientist*, p.53, 21 July 1988.
Tomorrow's World, Can We Do Without Nuclear Power?, BBC2, 19 November (1986).
UKAEA, Memorandum to the Royal Commission on Environmental Pollution, October 1974.
UKAEA, Environmental Impact Assessment etc., of the European Demonstration Fast Reactor Fuel Reprocessing Plant (EDRP) at Dounreay, Caithness (1985).
UK NIREX Ltd, The Way Forward, A Discussion Document (1987).

United Nations Scientific Committee on Effects of Atomic Radiation (UNSCEAR), *Sources and Effects of Ionising Radiation*, United Nations, (New York, 1977).
US Environmental Protection Agency, Air Quality Criteria for Particulate Matter and Sulphur Oxides (1981), Oxides of Nitrogen (1980), Polycyclic Organic Matter (1979), Arsenic (1980) and Lead (1977).
Wilkinson J., *Radioactive Waste Management, Technical Hazards and Public Acceptance*, p.199 (Oyez Scientific Technical Services, 1985).
Williams A., 'Economics of coronary artery bypass grafting', *Brit. Med. J.*, Vol.291, p.326 (1985).
Wilson R. and Jones W.L., *Energy, Ecology and the Environment* (New York, Academic Press, 1974).
Wrixon A.D., National Society for Clean Air, Scarborough Conference (1985).

GLOSSARY

Activation (Nuclear), The creation of radioactivity, usually in structural materials, by absorption of neutrons.

Adit, Sloping tunnel (leading down to the emplacement zone of a repository).

Advanced Thermal Reactor, AGR, PWR or SGHWR.

AGR, Advanced Gas-Cooled Reactor.

Alpha Radiation, Alpha particles (helium nuclei) ejected during radioactive decay of Heavy Atoms.

Atomic Number, The positive charge on the nucleus of an atom.

AVM, The French process for making HAL into glass, developed at Marcoule.

Becquerel (Bq), The unit of rate of radioactive decay, one disintegration per second.

Beta Radiation, beta particles (electrons) emitted during radioactive decay of some atomic nuclei.

Billion, The US billion or 1,000 million.

Biosphere, The part of the Earth in which living things are found.

BNES, British Nuclear Energy Society.

BNFL, British Nuclear Fuels Ltd.

CEGB, The Central Electricity Generating Board.

CHP, Combined Heat and Power.

CMS, Cemented Magnox swarf.

Criticality (with reference to neutrons in nuclear plants), the condition when fission is self-sustaining.

Critical Group, The group of the public most affected by a given operation.

Curie (Ci), The historical unit of radioactive decay rate, equivalent to 0.037 TBqs.

Daughter (in radioactive decay), the first decay product of a radionuclide.

Decay Constant, The fraction of atoms of a radionuclide which decay in unit time.

Decontamination, The removal of radioactive material from surfaces to which it has been transferred.

DCF, Discounted Cash Flow.

DOE, The UK Department of the Environment.

DoEn, The UK Department of Energy.

E, Defining a power of 10, e.g. 2E6 is 2 x (10 to the power 6) or 2 million.

EDF, Electricite de France, The French electricity supplier.

Enrichment (of isotopes), The increase in concentration of a chosen isotope of an element, e.g. uranium-235 in uranium.

ESI, The Electrical Supply Industry in England and Wales.

Evaporites, Inorganic salts left in geological deposits by the evaporation of natural waters.

Fast Reactor, A nuclear fission reactor in which the neutrons are not deliberately slowed down by a moderator.

FBC, Fluidised Bed Combustion (of coal).

FBR, A form of Fast Reactor in which net breeding of fissile isotopes (usually plutonium-239) occurs.

Fission (nuclear), The splitting of the nucleus of a Heavy Atom, usually by neutrons.

Fly Ash, The ash from the combustion of coal in power stations which 'flies', i.e. is entrained in the combustion gases.

FOE, The Friends of the Earth.

FRG, The Federal Republic of Germany.

Fusion (nuclear), The interaction of two atomic nuclei to form a larger nucleus.

Gamma Radiation, Electromagnetic radiation emitted during radioactive decay of atomic nuclei.

GDP, Gross Domestic Product.

Granddaughter (in radioactive decay), the second product of a decay chain, i.e. the daughter of a daughter radionuclide.

Greenpeace, A 'Green' organisation of strong anti-nuclear views.

GW(E), A gigawatt = 1000 megawatts or 1,000,000 kilowatts, E denoting electric power.

GW(E)yr, The energy transmitted in one year at one gigawatt.

H.A., Heavy Atoms, i.e. usually nuclei of mass over 200; in this book, uranium and nuclides derived from it by neutron reactions without fission.

HAL, High Active Liquors, i.e. the concentrates of the more highly active waste solutions during reprocessing.

Half-life, The period in which half the nuclei of a radioactive isotope decay.

HEW, Heat-Emitting Wastes, i.e. solid forms of the more highly active wastes which can cause thermal problems after burial.

HLW, High Level Wastes, usually vitrified forms of HAL and often synonymous with HEW.

HSC, Health and Safety Commission.

IAEA, The International Atomic Energy Agency.

IBA, The Independent Broadcasting Authority.

ICRP, The International Commission on Radiological Protection.

ILW, Intermediate Level Wastes, i.e. unshielded packages with over 4 Bq/milligram alpha or 12 Bq/milligram beta/gamma, but not HLW.

Isotopes, nuclei of the same element (or Atomic Number) with differing masses.

JET, The Joint European Torus, sited at Culham in Oxfordshire.

Kwhr, Electrical energy equivalent to 1 kilowatt for 1 hour

LLW, Low Level Wastes, i.e. waste with lower activity than ILW, but too high for conventional waste dumping.

Magnox, The magnesium/aluminium alloy used for canning metallic uranium in Magnox reactors.

Mass Number, The number of neutrons plus protons in an atomic nucleus.

Matrix (for waste), A solid, often cement, in which radioactive waste is embedded.

Micron, A micrometre or one thousandth of a millimetre.

Millennium, One thousand years.

Mine Spoil, The residue after removing usable coal from coal as mined.

Moderator, The material in a Thermal Reactor which slows down neutrons by elastic (non-absorbing) collisions.

MOX, Mixed Oxide Fuel, i.e. a nuclear fuel containing a mixture of plutonium and uranium oxides.

MW(E), A megawatt = 1,000 kilowatts, E denoting electric power.

MW(Th), A megawatt = 1,000 kilowatts, Th denoting thermal output, often connected with the heat rating of fuel elements.

MWD(Th), The thermal energy equivalent to a rating of 1,000 kilowatts for one day, often describing the irradiation of fuel.

NEL, The National Engineering Laboratory.

Neutron Number, The number of neutrons in an atomic nucleus.

NHS, The National Health Service.

NIH, An acronym for Not Invented Here.

NII, The Nuclear Installations Inspectorate.

NIMBY, An acronym for Not In My Backyard.

NIREX, The Nuclear Industry Radioactive Waste Executive, now UK NIREX Ltd.

NNC, The National Nuclear Corporation.

NRPB, The National Radiological Protection Board.

NTC, Nuclear Technology (Consultants) Ltd.

NUM, The National Union of Mineworkers.

PCM, Plutonium-Contaminated Materials (wastes).

PFBC, Pressurised Fluidised Bed Combustion (of coal).

Photovoltaic Cell, a unit converting light energy directly into electricity.

PuE, The Equivalent Quantity of Pu239, in terms of reactivity in a Fast Reactor.

PV, Present Value.

PWR, The Pressurised Water Reactor, developed in the US by Westinghouse.

Radionuclide, A radioactive isotope of an element.

RCEP, The Royal Commission on Environmental Pollution.

Rem, The historical unit of radiation impact on man, equivalent to 0.01 Sieverts.

'Renewable' Power Systems, Sources of power such as wind and wave which are periodically created on Earth, primarily by the Sun and Moon.

SEA, Sea Energy Associates, A group developing the 'Clam' system of wave power.

SGHWR, The Steam Generating Heavy Water Reactor, a prototype pressure tube reactor operating at Winfrith Heath, Dorset.

Spent Fuel, The fuel discharged from a reactor when its reactivity is no longer sufficient.

SSEB, The South of Scotland Electricity Board.

Stochastic, Random, as in the occurrence of cancers after cell damage.

Sv, sievert, The current unit of radiation dose to the human body (elaborated in Section 5.4.2).

TBq, Terabecquerel = one million million Becquerels or 27 curies.

Thermal Reactor, A nuclear fission reactor in which the neutrons are deliberately slowed down by a moderator.

THORP, The Thermal Oxide Reprocessing Plant, i.e. the plant for reprocessing oxide fuel from Advanced Thermal Reactors.

Tonne, The metric ton or 1,000 kilograms, very nearly the same as the conventional ton used commonly in the UK.

UDM, The Union of Democratic Mineworkers.

UKAEA, The United Kingdom Atomic Energy Authority.

USL, Under Sea-bed from Land.

WHO, World Health Organization.

INDEX